T0313346

Digital Innovation and the Future of Work

RIVER PUBLISHERS SERIES IN INFORMATION SCIENCE AND TECHNOLOGY

Series Editors:

K. C. Chen
National Taiwan University, Taipei, Taiwan
and
University of South Florida, USA

Sandeep Shukla
Virginia Tech, USA
and
Indian Institute of Technology Kanpur, India

Indexing: All books published in this series are submitted to the Web of Science Book Citation Index (BkCI), to SCOPUS, to CrossRef and to Google Scholar for evaluation and indexing.

The "River Publishers Series in Information Science and Technology" covers research which ushers the 21st Century into an Internet and multimedia era. Multimedia means the theory and application of filtering, coding, estimating, analyzing, detecting and recognizing, synthesizing, classifying, recording, and reproducing signals by digital and/or analog devices or techniques, while the scope of "signal" includes audio, video, speech, image, musical, multimedia, data/content, geophysical, sonar/radar, bio/medical, sensation, etc. Networking suggests transportation of such multimedia contents among nodes in communication and/or computer networks, to facilitate the ultimate Internet.

Theory, technologies, protocols and standards, applications/services, practice and implementation of wired/wireless networking are all within the scope of this series. Based on network and communication science, we further extend the scope for 21st Century life through the knowledge in robotics, machine learning, embedded systems, cognitive science, pattern recognition, quantum/biological/molecular computation and information processing, biology, ecology, social science and economics, user behaviors and interface, and applications to health and society advance.

Books published in the series include research monographs, edited volumes, handbooks and textbooks. The books provide professionals, researchers, educators, and advanced students in the field with an invaluable insight into the latest research and developments.

Topics covered in the series include, but are by no means restricted to the following:

- Communication/Computer Networking Technologies and Applications
- Queuing Theory
- Optimization
- Operation Research
- Stochastic Processes
- Information Theory
- Multimedia/Speech/Video Processing
- Computation and Information Processing
- Machine Intelligence
- Cognitive Science and Brian Science
- Embedded Systems
- Computer Architectures
- Reconfigurable Computing
- Cyber Security

For a list of other books in this series, visit www.riverpublishers.com

Digital Innovation and the Future of Work

Editors

Hans Schaffers

Adventure Research, The Netherlands

Matti Vartiainen

Aalto University, Finland

Jacques Bus

Digital Enlightenment Forum, The Netherlands

LONDON AND NEW YORK

Published 2020 by River Publishers
River Publishers
Alsbjergvej 10, 9260 Gistrup, Denmark
www.riverpublishers.com

Distributed exclusively by Routledge
4 Park Square, Milton Park, Abingdon, Oxon OX14 4RN
605 Third Avenue, New York, NY 10017, USA

Digital Innovation and the Future of Work / by Hans Schaffers, Matti Vartiainen, Jacques Bus.

Routledge is an imprint of the Taylor & Francis Group, an informa business

ISBN 978-87-7022-220-4 (print)

While every effort is made to provide dependable information, the publisher, authors, and editors cannot be held responsible for any errors or omissions.

Contents

8 Game-changing Technologies: Impact on Job Quality, Employment, and Social Dialogue 157

Eleonora Peruffo and Enrique Fernández-Macías

9 The Diversity of Platform Work—Variations in Employment and Working Conditions 177

Irene Mandl and Cristiano Codagnone

10 Workplace Innovation and Industry 4.0: Creating Synergies between Human and Digital Potential 197

Peter Totterdill

Preface

Work is a fundamental human activity and forms an important part of our lives. For many people, work is a goal-oriented activity, allowing us to create something useful or, as in the arts, beautiful. Work also enables us to learn and improve skills and capabilities, shape our creativity and fulfil our human potential. However, working as an activity can take many different forms, including hard labor under difficult circumstances aimed only at earning a living. This can be observed in low-quality jobs in developing countries but also in the 'gig' jobs of the modern platform economy. Technology, and in particular those set of technologies that drive the ongoing process of digitalization, plays a key role in shaping the modern world of work. It does so in various ways, for example by enabling new forms of team cooperation and working on a distance, by influencing how work activities are organized and coordinated in different ways such as in customer driven supply chains, networks or platforms, by assisting the worker and teams in conducting complex tasks, by automating work tasks, and by monitoring and managing working behavior and performance. Digital technological innovations and the drive towards new business models exploiting such technologies are bringing important work-related issues on the forefront, such as quality of work, working conditions, worker participation, self-organization of work, and workplace innovation. Digital innovations also raise broader societal and ethical debates regarding data surveillance, data sovereignty, privacy intrusion and market dominance. Technologies that constitute the current wave of digitalization, such as artificial intelligence, Internet of Things, robotics, big data, wireless networks, enterprise platforms and other will further influence on transforming the nature of work, the work environment and its societal context. Therefore, it is important to understand the implications of digitalization for the future of work and the future work environment from a human and societal perspective. To this end, this book brings together a collection of studies from different perspectives and authored by a diverse group of experts to address the technological, economic and political forces shaping the new world of work and the prospects for human-centric and responsible innovations. Five major topics are covered: 1. The evolution of digital technology impacting ways of working; 2. The role of artificial intelligence in new ways of working; 3. Transformation of work, jobs and employment; 4. Digitalization and need for skills and competencies; and 5. New forms of decentralized working and cooperation. We hope the book will be useful not only for scientists, engineers and students but also for practitioners and policy makers.

Hans Schaffers, The Netherlands
Matti Vartiainen, Finland
Jacques Bus, The Netherlands

List of Contributors

Abhishek Gupta, Montreal AI Ethics Institute, and Microsoft, Canada

Christian Korunka, Department of Work and Organisational Psychology, Faculty of Psychology, University of Vienna, Vienna, Austria

Christiano Codagnone, University of Milan, Milan, Italy

Eleonora Peruffo, European Foundation for the Improvement of Living and Working Conditions (Eurofound), Dublin, Ireland

Emilio Mordini, Responsible Technology SAS, Paris, France, and Health and Risk Communication Center, University of Haifa, Haifa, Israel

Enrique Fernández-Macías, Joint Research Centre of the European Commission, Seville, Spain

Frans van der Zee, TNO, Netherlands Organisation for Applied Scientific Research, Leiden, The Netherlands

Hans Schaffers, Adventure Research and Radboud University, The Netherlands

Irene Mandl, European Foundation for the Improvement of Living and Working Conditions (Eurofound), Dublin, Ireland

Jacques Bus, Digital Enlightenment Forum, The Netherlands

Jutta Treviranus, Inclusive Design Research Centre, Ontario College of Art and Design University, Toronto, Canada

Karolus Kraan, TNO, Netherlands Organisation for Applied Scientific Research, Leiden, The Netherlands

Matti Vartiainen, Department of Industrial Engineering and Management, Aalto University, Espoo, Finland

Markku Wilenius, University of Turku, Turku, Finland

Michel Bauwens, P2P Foundation, Amsterdam, The Netherlands

Osmo Kuusi, University of Turku, Turku, Finland

Paul Preenen, TNO, Netherlands Organisation for Applied Scientific Research, Leiden, The Netherlands

Peter R.A. Oeij, TNO, Netherlands Organisation for Applied Scientific Research, Leiden, The Netherlands

Peter Totterdill, Workplace Innovation Europe, Cork, Ireland, and Glasgow, United Kingdom

Risto Linturi, R. Linturi Plc, Finland

Sarah Manski, George Mason University, Fairfax, VA, USA

Sofi Kurki, University of Turku, Turku, Finland

Steven Dhondt, KU Leuven, Leuven, Belgium; and TNO, Netherlands Organisation for Applied Scientific Research, Leiden, The Netherlands

Valerie Frissen, eLaw – Center for Law and Digital Technologies, Leiden University, and SIDN Fund, The Netherlands

List of Figures

List of Tables

List of Abbreviations

AI	Artificial Intelligence
AM	Additive Manufacturing
ARPA	Advanced Research Projects Agency
AR/VR	Augmented Reality / Virtual Reality
ART	Anticipated Radical Technology
AS	Autonomous Systems
CAD/CAM	Computer Aided Design/Computer Aided Manufacturing
CBPP	Commons-Based Peer Production
CERN	Conseil Européen pour la Recherche Nucléaire (European Organization for Nuclear Research)
COVID-19	Pandemic caused by coronavirus (SARS-COV-2)
DAI	Distributed Artificial Intelligence
DevOps	Practices combining (Software) Development and (IT) Operations
DLT	Distributed Ledger Technologies
EHR	Electronic Hospital Records
EPR	Electronic Patient Records
ERP	Enterprise Resource Planning
EV	Electric Vehicles
EU	European Union
EU-KLEMS	Database for EU-level analysis of Capital, Labour, Energy, Manufacturing, Services
EUWIN	European Workplace Innovation Network
FLOSS	Free/Libre and Open Source Software
GCT	Game-Changing Technologies
GDPR	General Data Protection regulation
GNP	Gross National Product
GOFAI	Good Old-Fashioned Artificial Intelligence
GVN	Global Value Producing Network
HIV	Human Immunodeficiency Virus
HLEG	High Level Expert Group

IC	Intensive Care
ICO	Initial Coin Offering
ICT	Information and Communication Technologies
ID	Identity Document
IP	Internet Protocol
ILO	International Labour Organization
IoT	Internet of Things
KBC	Knowledge Based Capital
ML	Machine Learning
MLM	Mobile Labor Markets
MR	Mobile Robotics
MS	Microsoft
OECD	Organisation for Economic Cooperation and Development
OODD	Out-Of-Distribution Data
PIAAC	Programme for the International Assessment of Adult Competencies
PoC	Proof of Concept
P2P	Peer-to-Peer
RBTC	Routine Biased Technical Change
R&D	Research and Development
RTI	Radical Technology Inquirer method
SME	Small and Medium-sized Enterprises
SMS	Short message Service
STEM	Science, Technology, Engineering, and Mathematics (competencies)
STSD	Socio-Technical Systems Design
TQM	Total Quality Management
VI	Vertical Integration
VT	Virtual Team

1

Introduction

Hans Schaffers[1], Matti Vartiainen[2], and Jacques Bus[3]

[1] Adventure Research, The Netherlands
[2] Aalto University, Espoo, Finland
[3] Digital Enlightenment Forum, The Netherlands

1.1 Digitalization in a Changing Society

The concept of digitalization captures the widespread adoption and impact of digital technologies in our lives, in our workplaces, organizations and business models, and in the transformation of our economy and society. Digital technologies for data processing and communication underly far-reaching and high-impact innovations such as the Internet of Things, artificial intelligence (AI), big data and data analytics, robotics, wireless multimedia, enterprise platforms, social networks, and blockchain. These digital innovations embody important new opportunities for prosperity and well-being. There is no shortage of publications explaining how digital technologies and their applications enable innovative products, services and business models, and how they facilitate new networked forms and patterns of collaboration, production, and working. Visionary scenarios such as Industry 4.0 and smart cities testify to a technology-optimistic view. While the process of innovation, experimentation, and practical adoption is ongoing and alternating successes with failures, such innovations are transforming a range of sectors including manufacturing, health care, transport and logistics, food and retail, energy, and public and commercial services. Additionally, digital technologies and applications bring considerable changes in job quality, working conditions and work practices, organization and management, and in the very nature of labor and work as a human activity. This way they deeply affect people's activities, behaviors, and daily lives. In terms of the changes induced by digital innovations and accelerated by the network effects they are often subject to, they can be considered as game-changing, pervasive, and disruptive, in multiple ways.

As digital innovations often require integration in a broader technical, organizational, and social environment, they cannot only be considered through the lens of purely technical innovations as represented by electronic systems, networking technologies, software infrastructures, and networked applications. Digital innovations relevant for the future working environment must contribute to and form part of a more comprehensive sociotechnical systemic innovation acceptable to workers. Examples include platform-based work systems, value creating networks, and urban public service systems. The creation and acceptance of such systemic workplace-related innovations require the alignment of a diversity of technologies, processes, organizations and people, as well as stakeholder engagement, and capabilities for adaptation, integration, and learning.

An important aspect of digital technologies and their applications in an increasingly connected world is how they are offering new opportunities for collecting and analyzing all kinds of data, including personal data, and for control and coordination based on such data, while respecting privacy and other ethical considerations. A data-driven approach allows a more efficient and effective organizational coordination, cooperation, and networking, for example in often complex and globalized supply networks, as well as the coordination of work activities in such networks. But we should not forget the potential negative aspects of such globalized supply chains, in terms of commodification of work, lacking worker rights and protections, and unregulated exploitation of natural resources, which creates wide ranging negative external effects. Given these effects and possibly accelerated by the current Corona pandemic, we may enter a period of more seriously considering pathways toward alternative, more decentralized and localized production and work systems, which give more attention to workers' satisfaction.

The way digital technological innovations are created and applied is often the focus of intense societal debates. Key issues include privacy and security; surveillance, control, and power; quality of work and work-life balance; competition and market dominance; and digital divides and social cohesion. In relation to these debates, our society is facing a range of often complex dilemmas and different interests in the continuing process of creating, deploying, and applying digital technological opportunities. At the societal level, the strive for economic benefits of digitalization sometimes hinders social innovation and cohesion. As extensively studied by

Soshana Zuboff in "The Age of Surveillance Capitalism" [1], surveillance-based business models based on commodifying personal data, such as created by Google and Facebook, are conflicting with the idea of data sovereignty and self-determination. At the level of business, ethical dilemmas emerge regarding the exploitation of employee data, which can create conflicts with respect to responsible employee policies. The identification and understanding of the dilemmas and challenges surrounding digital innovation and the stimulation of the public discourse is already one of the challenges in how society addresses and is giving shape to digital innovation. Technological impact analysis, development of digital ethics principles, implementing corporate social responsibility policies, creating responsible and user-centric forms of innovation, and stimulating the public debate are among the wide variety of necessary but also inherently imperfect ways of handling these challenges. A prerequisite for resolution of these dilemmas is the public debate on sometimes difficult trade-offs between different interests and diverging worldviews involved.

Against this background, this book focuses on understanding the implications of "digital innovation" in relation to "the future of work." Thomas Malone's book "The Future of Work" [2] took its point of departure in the role of information and communication technologies in enabling new forms of decentralized organization such as networks, internal markets, and communities. Yochai Benkler's "The Wealth of Networks" [3] elaborated on the model of peer production and sharing, and on the economics of social production (the contribution by Michel Bauwens and Sarah Manski in this book is in addressing the related topic of commons-based peer-to-peer economies). Another key development has been the combined development of AI, platform ecosystems, and crowd-based forms of organizing [4]. Clearly all this has important implications with respect to the "world of work," which includes besides the changing nature of work also the new forms of digitally enabled organizing and cooperation. This book aims at exploring these implications in more detail including the ways for handling the actual and potential dilemmas and challenges involved. In this sense, the book takes a human- and society-centric perspective regarding the emerging practices of digital working, cooperating, and organizing. Our intention is to shed light on the forces shaping the new "world of work" and on the prospects for "human-centric" and "responsible" digital innovation in that context. To this end, the book critically examines and identifies the actual and potential impacts, implications, and challenges of digital innovation as related to work. Let us now first discuss how the nature of work is transforming due to the forces of digitalization.

1.2 The Future of Work Revisited

1.2.1 Mechanisms that Transform Ways of Working

Converting information to a digital format, digitalization of tools, products and services, value-adding processes, working environments, and the adoption of digital business models gradually change the nature of work and ways of working in micro-, small-, medium-sized, and large companies, in local regions such as cities, and globally on virtual online platforms. This development is shown as changes in the work settings or work environment, and where and when work is done. Digitalization penetrates the components of sociotechnical work systems and their contexts to varying degrees, generating the need to update human practices and competencies.

The technology affects a work process by transforming the microstructure of its phases, which is then mirrored in other components of the work system. As the value-producing process is usually still steered and managed by human activities, the changes are also needed in human tasks and jobs. Through this process of transformation, digitalization also replaces, hybridizes, and renews tasks and job contents. This, on its behalf, leads to a need to reorganize and structure organizations. As a consequence, many other organizational and human-related issues are under pressure. For example, restructuring and organizing of work units are needed, resulting in new ways of working, leadership practices, and competencies. The outcomes of this transformation are digitally influenced products and services. The impacts of digitalization do not stop here, as they reflect on how to organize and manage work relations. The remaining hybrid and newly created tasks and jobs need restructuring and reorganizing.

The COVID-19 pandemic forced us to experiment with transformed ways of working in practice. A survey by Eurofound [5] shows that over a third (37%) of those working in the European Union began to telework as a result of the pandemic—over 30% in most member states. The largest proportions of workers who switched to working from home were found in the Nordic and Benelux countries (close to 60% in Finland and above 50% in Luxembourg, the Netherlands, Belgium, and Denmark, and 40% or more in Ireland, Sweden, Austria, and Italy). For comparison, a present US survey study [6] collecting a total of 25,000 responses from April 1 to

April 5, 2020, showed that over one-third of the workers responded to the pandemic by shifting to remote work, while another 11% were laid off or furloughed [7]. Younger people were more likely than older people to switch from commuting to remote work. A representative survey among the German working population [7] tells that every second one was working fully or partly from home in mid-March 2020. Meanwhile, 41% responded that their job is not suitable for working from home. The studies on telework during the pandemic show that most teleworkers collaborated virtually with their colleagues, managers, and customers during their working days. Therefore, digital competences are especially needed in remote virtual work mode. These and similar studies around the globe tell that this "natural experiment" has until now brought forth unanswered questions on how to anticipate these kinds of partly unexpected situations, organize remote work and working conditions, and needed social and virtual support.

1.2.2 Future New Ways of Working Based on "Old" New Ways of Working

What is common to new ways of working is "flexibility." The *flexible structures* of working life involve organizing assignments into temporary projects for which a group of workers is assembled as necessary to realize the project at hand. Virtual dispersed teams are used as a standard way of cooperating. The main trend is to build flexible, adaptable forms of organizations. More often, work is organized within and between workplaces into *temporary projects*. Also, individual assignments are often temporary, especially when working on platforms.

At the individual level, different forms of remote work lay a basis for other ways of working. In more detail, when work assignments are done outside the "main place of work," it is referred to as remote work or telework. Thus, remote work is defined as "working outside the conventional office using telecommunication-related technologies to interact with supervisors, coworkers, and clients." When employees are changing their working locations daily or weekly, they are doing mobile and multilocational work. The development of digital working environments, such as global online platforms employing microproviders, has expanded working locations worldwide. Fully digital work has appeared detached from any stabilized social and organizational settings as *detached temporary global telework*. Detached global teleworkers often work from their homes on different continents.

Remote workers often collaborate from their present locations in temporary or permanent virtual teams and communities. The second main type of the now traditional "new ways of working" is virtual collaboration in distributed teams and projects. Distributed or dispersed work also has deep historical roots. Team members working in different locations and their geographical distances from each other constitute a distributed team. These teams become virtual when extensively utilizing various forms of computer-mediated communication that enable geographically dispersed members to coordinate their individual efforts and inputs.

Digital online platforms serve more and more as crowdsourced dispersed workplaces. A "platform" is a "palette" consisting of usable components to be used for different purposes by one or several actors. Platforms as the working context can be global and/or local. Such work platforms (e.g., Amazon Mechanical Turk) act as employment agencies. An employer somewhere in the world digitally posts tasks for the site's users to complete. A worker enters the platform using his/her own or a borrowed device, selects a task, completes it, gets credited with the proceeds, and selects the next task. Each completed task earns the worker remuneration.

A key issue for the future organizations is to become more resilient. Resilience does not only refer to adaptation, but rather to the capability for renewal of ways of working. In the future, it is necessary to be ready to absorb external shocks and to rapidly learn from them, while simultaneously anticipating new ones. Resilient people; work processes; work environments; and organizations get knocked down and get up again, ready to learn from events, and to be ready for future challenges. In this sentence I am not sure if semicolons should be used or rather commas.

1.3 Artificial Intelligence and Working Life

AI is an example of key technologies studied in several chapters of this book creating considerable dilemmas for society in general and future of work in particular. The scope of potential applications of AI in the production and work environment is wide ranging. Think of self-learning robots, quality control and predictive maintenance in manufacturing, the use of big data in supply chain management, and the use of AI in decision support in marketing, sales, and human resource management. There is also an increasing attention toward potentially negative aspects of AI, such as its impact on job quality, surveillance, and decision-making. In the beginning of June 2020, Amazon, IBM, and Microsoft decided to put the brake on the development of face recognition, which is based on machine learning AI. It appeared that in the current state of affairs, it only worked properly on white men, and, for example, very badly on black women. In a letter to the Congress published on the IBM's website, IBM's CEO Arvind Krishna said: "because there is the danger that this technology will be abused for massive surveillance and racial profiling, in particular law enforcement should limit its use" [8]. Of course, this decision was made against the background of the "Black Lives Matter" movement that started in the United States. A good thing is that with this decision a myth about big data was dumped. It is indeed not true that more data used for machine learning automatically leads to more precise AI predictions, as the data need to be representative for the group for which the results are used. This is addressed in Chapter 5 of this book.

A second myth, also discussed in Chapter 5, concerns the basic methods for machine learning predictive analysis. Predictions, even if a large and good representative data set is used for machine learning of the algorithm, will be based on statistical averages and a limited set of behavioral parameters. In particular, predicting human behavior based on the average behavior of a representative group of humans does not take into account that humans are not behaving averagely and worse, they can behave very irrationally. But during machine learning, outliers will be discarded or hardly given considerations. The consequence is that although a certain behavior can be likely, it cannot be taken for true, certainly not if the person under consideration is himself/herself an outlier (e.g., a person with a handicap).

A third myth is that machines will become more powerful than people and will therefore take over the world and make human beings slaves. This is discussed in detail in Chapter 4 of this book. The conclusion of Emilio Mordini, who has written this chapter, is "that AI is great as far as it is used as a tool to amplify and enhance human analytic, dichotomic skills. It is not good for understanding emotions and enjoying art expressions. The current debate about the hypothetical risk that AI might one day surpass human intelligence is largely misplaced."

Nevertheless, serious dangers exist in using AI, as appeared from the above-mentioned decision of three companies on face recognition. Human beings crave to belong and be accepted by their group, society, or nation. This characteristic can be used by people in power to force them through strong surveillance in a harness being designed for them based on AI predictions, e.g., through a "good citizen rating system" rewarding good behavior. Such systems are already under development and in use, for example, in China. Those companies in democratic states recognize this is laudable. But whether it is enough to avoid this is questionable. The question whether the machines surpass human intelligence is likely not very opportune, but whether people will be made to act like machines by an intelligent and powerful few, is probably a realistic danger. Of course, this does not hold in the political context only. It also is applicable to the working environment, and maybe even more so, given the dependencies between an employer and an employee. In this sense, this book offers some fundamental insights into the future of work.

1.4 Digital Work and the Human Condition

The increasingly important role of digital technologies such as AI in combination with current societal challenges due to the Corona pandemic and climate change raises new questions regarding the "future of work" in a society characterized by connectivity and systemic interdependencies. In discussing "work" in his recent book, "21 Lessons for the 21st Century," Yuval Harari [9] expects a decrease in jobs due to merging of information technology, robotics, and biotech. Learning algorithms and biometric sensors enable computers to be continuously updated and easily integrated in networks, which may considerably improve health care, for example, in developing countries. In these and other sectors, jobs may disappear due to the applicability of AI, causing the possibility of mass unemployment, social disruption, and even shifting dictatorial power to algorithms. In a recent article [10], Harari reflects on the "world after coronavirus" and suspects that the current emergency measures may become a "fixture for life," as some countries have implemented far-reaching forms of data surveillance to monitor the spread of coronavirus.

This current situation also invites to reflect about what philosopher Hannah Arendt called the "human condition." In her still important book of more than 60 years ago under this title, Arendt designates three fundamental human activities in the *vita activa*: "labor," "work," and "action" [11]. Throughout human history these activities have changed, shaped by the same conditions they helped creating them. What we currently call "working life" may combine these elements. In her vision, *labor* aims at fulfilling the direct demands and bare necessities of life. Arendt understands the human condition of labor as life itself, which necessitates labor, which is cyclic and repetitive and aimed at consumption. One may think of not only low-quality jobs in developing countries but also the flexible low-quality jobs in the platform economy. Distinct from labor, *work* aims at creating a world of useful and durable things, and is associated with efficiency, utility, and applying technology to exploit nature and become its master. At the same time, this attitude, in the extreme form which we are witnessing in the current era called the Anthropocene, creates irreversible destruction of our living environment with potentially grave consequences for current and future generations. Maybe it also brings the insight that after all human beings are not disconnected from nature and that we are part of a world characterized by all kinds of interdependencies. Arendt's third activity of *action*, which also includes *speech*, relates us to other humans in a "web of relationships" and is the domain of plurality and politics. In this sense, Arendt invites us, in the current age of digitalization and global societal challenges, to rethink the "world of work."

1.5 Overview of Book Contributions

Whereas all contributions address the future of work, cooperation, and organization in a context of digitalization, there is a broad diversity of perspectives. The initial chapters address the long-term evolution of (digital) technology in relation to future working challenges. The next set of chapters select their starting point in studying the implications of AI on humans, work, and organizations. Thereafter, a third collection of chapters addresses various aspects of the future of work and employment as transformed by the forces of digitalization. The fourth theme is how digital technologies affect human capabilities, skills, and competences. Finally, several chapters focus on how digital technologies are enabling new forms of cooperation and innovation in open and decentralized networks.

1.5.1 Technology Evolution and Future Ways of Working

In Chapter 2, Sofi Kurki and Markku Wilenius take their point of departure in the expected impact of the Coronavirus and similar on the future of work and employment. The authors explore the future of work through two different long wave theories, implying that human sociotechnical systems can be understood in terms of long range cyclical renewal processes. While the well known Kondratieff wave theory addresses the long-term transformation of socioeconomic systems, the more recent but lesser known approach of Malaska focuses on socioevolutionary theory of social change. Both streams of work raise questions regarding the transformation of work and evolution of skills in human-centric organizations. The authors conclude—also motivated by the implications of the current Corona pandemic for the nature of work and employment—that there is an even greater need to focus on human skills and capabilities that make us better as human beings.

A detailed assessment of the impact of emerging technologies on society and, in particular, working life is presented by Risto Linturi and Osmo Kuusi in Chapter 3. The authors describe a framework for technology foresight and technology driven societal transformation presenting recent findings of a foresight research program, which is still ongoing. In using transition theory concepts for integrating technology push and demand pull, the foresight study addresses the linkages between 100 radical technologies and 20 types of societal and human goals associated with the so-called "global value-producing networks," such as passenger transport, manufacturing, built environment, and automation of work. The authors demonstrate that all of such value-producing networks face a paradigm-level transformation due to anticipated progress of enabling radical technologies and goal-seeking behavior of actors in each network, which will challenge existing professions, the way we organize work, use our tools and capabilities, and even the need to work for money.

1.5.2 Artificial Intelligence, Humans, and Work

Chapter 4, authored by Emilio Mordini, offers fundamental insights into the relation between human and artificial intelligence. The author observes that our epoch is fascinated by human-like minds and by the fantasy of intelligent machines, which might surpass and substitute human intelligence. Intelligent machines may imitate human skills, such as intuition, emotions, sentiments, capacity for perceiving atmosphere and context, and human's capacity for sense of humor. The author argues that computational machines may replicate human intuitive skills but they cannot exactly duplicate them. The chapter neither aims to grade natural vs. artificial intelligence nor to raise any ethical consideration on AI; rather aims to show the inherent limits of AI and its applications. A conclusion is that the current debate about the hypothetical risk that AI might one day surpass human intelligence is largely misplaced.

A more design-oriented perspective is offered by Abhishek Gupta and Jutta Treviranus in Chapter 5. They argue that AI can either automate and amplify existing biases or provide new opportunities for previously marginalized individuals and groups. Small minorities and outliers are frequently excluded or misrepresented in population data sets. Even if their data is included, data-driven decisions favor the statistical average, thereby disadvantaging small minorities. Small minorities and people at the margins are also most vulnerable to data abuse and misuse. The authors state that current privacy protections are ineffective if you are an outlier or in some way anomalous. Their chapter discusses the challenges, dangers, and opportunities of machine learning and AI for individuals and groups that are not represented by the majority.

Valerie Frissen in Chapter 6 argues that AI has a huge potential to facilitate, enhance, and transform human activities. However, concerns have arisen about the risks involved, and a strong call for new ethical and regulatory frameworks has emerged helping to build human-centered, responsible approaches to the use of AI. The chapter discusses two cases of using ethical guidelines related to AI in working practices. The first case addresses approaches to develop responsible AI in health care, more specifically for intensive care, whereas the second case focuses on using AI for data-driven approaches in the domain of public safety and organized crime. The author observes that in both cases particularly the issue of data quality raises concerns. A conclusion is that rigorous and standardized protocols will be required for collecting and using data for AI applications. Furthermore, the two cases demonstrate interesting differences and practical complexities in the way moral and ethical considerations are being taken into account.

1.5.3 Transformation of Work and Employment

The way digital technologies affect the quality of work is addressed by Christian Korunka in Chapter 7. First, a short overview of the general development of digital technologies in the "world of work" is presented, focusing on impacts of "conventional" digital technologies such as computers, office equipment, and software tools. The author then discusses theoretical approaches aimed at explaining these effects, based on the social acceleration concept of Hartmut Rosa and the concept of paradoxes in digital technologies. Based on these concepts, new demands related to digital technologies in the current world of work are discussed and examples of empirical studies investigating these new demands are presented. The author proposes several evidence-based recommendations with the aim of increasing quality of working life when working with digital tools.

A closer look into the work-related implications of eight game-changing technologies is offered by Eleonora Peruffo and Enrique Fernández-Macías. In Chapter 8, these authors argue that since the industrial revolution, economic development has been punctuated by leaps driven by the successive introduction of radical technological breakthroughs. Until recently, these breakthroughs mostly concerned the manufacturing sector, with services acting as a kind of residual category that collected all the labor displaced by technological progress. The digital revolution has led to the emergence of several major technological breakthroughs with a very significant disruptive potential for both manufacturing and services in Europe. The authors bring together qualitative research findings spanning from 2015 to 2018 which describe the potential impacts of eight different game-changing technologies on work and employment in Europe. The technologies studied increase flexibility and allow mass customization of goods and services, make manufacturing and services increasingly undistinguishable, and increase the efficiency and control of economic processes. In addition, these technologies often have labor-saving effects that may transform the structure of employment and tasks in Europe. The authors conclude that negative implications in terms of autonomy, privacy and control of workers, and conflicts related to the opacity of managerial algorithms or the ownership of the data generated in the workplace will become increasingly salient.

The role of platform work is the focus of Irene Mandl and Christiano Codagnone in Chapter 9. Platform work has emerged as an employment form and business model in Europe about a decade ago, and while still small in scale, it is dynamically developing. This also refers to an increasing heterogeneity within platform work, which results in different effects on employment and working conditions of platform workers. The authors propose a classification of platform work using a combination of five criteria: (1) scale of tasks, (2) skills level required to fulfill them, (3) format of service provision, (4) selector of task assignment, and (5) form of matching. The classification identifies 10 distinctive types of platform work, and the chapter discusses the employment and working conditions of platform workers affiliated to 5 of these 10 types. The authors stress that there is no type of platform work, which exclusively poses advantages or disadvantages to the workers, and that their opportunities and risks vary quite substantially. The authors conclude that platform work, which is related to small-scale, low-skilled tasks (algorithmically) assigned to the worker by the platform which— beyond matching—also determines work organization tends to raise more challenges for workers and the labor market. The chapter finally proposes a differentiated policy approach which better considers the heterogeneity in platform work.

1.5.4 Digitalization and the Need for Skills and Competencies

Peter Totterdill, in Chapter 10, discusses the need for aligning human capabilities and digital opportunities of Industry 4.0. He argues that Industry 4.0 is at risk of being no more than the latest in the long line of technological predictions based on exaggerated claims. The risk is of drawing corporate decision-makers into patterns of investment that ultimately fail because they ignore the required synergy between the design and implementation of technologies, on the one hand, and human and organizational factors on the other. The technological advances represented by Industry 4.0 potentially offer real economic and also social benefits. At the same time, realizing this potential and avoiding the mistakes of the past means recognizing the importance of a new and more inclusive paradigm of innovation. The challenge is that of reconciling the ordered, rational organization of work offered by emergent technologies with the creative, dialogical, serendipitous, and even chaotic human interactions that can stimulate innovation.

A closer look at human competencies and skills in digitalized working environments is offered by Matti Vartiainen in Chapter 11. He argues that decisions to integrate digital technologies into work processes have a wide influence in work organizations and their currently used and in the future required competencies. The chapter concentrates on exploring how digitalization is related to present and future competencies by transforming work processes, task and job contents, and the organization of labor, and finally products and services. Changes in work processes produce changes in job and task structures as jobs and tasks are replaced and destroyed when human labor is removed, hybridized when new features and demands are added, and recreated when new, previously unseen work requirements emerge. This, in turn, creates the need for reorganization. This way, digitalization increases pressures to rearrange work system elements anew. In addition to new ways of working and leadership, new competencies are needed. Future-oriented competencies are needed at the individual, organizational, and societal levels. They include competencies to adapt and create new ways of working, anticipation, and digital competencies.

Steven Dhondt, Frans van der Zee, Paul Preenen, Karolus Kraan, and Peter Oeij in Chapter 12 describe a new approach to investigate, unravel, and explain the implications of digital technologies for skills. To do so, the chapter develops an approach to assess technology in companies in a more precise way, building on three main arguments. Firstly, current approaches to the subject treat all (new and emerging) technologies as equal. A more specific approach to technology is needed. Secondly, instead of starting from the potential of digital technologies, the focus should be on how technology investment decisions of companies are actually taken. Companies do not automatically reason from the available technology potential, but rather build on their current technology and capital stock and competitive position (the potential of technology). Thirdly, the organizational context should be considered. The actual use of skills in companies is strongly related to the organizational context. This is identified as the dominant organizational context. Based on these three main arguments, the authors propose a new research framework for technological impact on work and skills which is applied to two professions in Dutch industry.

1.5.5 New Forms of Decentralized Working and Cooperation

The focus of Chapter 13, authored by Hans Schaffers, is on the changing innovation process in an increasingly digitalized and networked working environment. Products, services, business processes, workplaces, and supply chains become connected, enabled by digital technologies and platforms, and result in new ways of value creation based on new types of business models and new forms of cooperation and governance of innovation. Digital innovations also affect the public sector and the nature of public services, for example, in health care, energy, education, transport, and urban life in the "smart city." Digitalization reshapes the process of innovation as the combined result of three changes: (a) the changing nature of innovations as digital artefacts, (b) the changing networked and platform-based environment of innovation, and (c) the changing rules and roles in innovation governance. The chapter discusses characteristics of the new forms of innovation in industry, the public sector, and in decentralized ecosystems.

Finally, Michel Bauwens and Sarah Manski in Chapter 14 discuss a new model of value creation called "commons-based peer production" (CBPP). The authors discuss whether it offers new solutions for integrating externalities in our economic systems. CBPP are open, collaborative ecosystems that allow for a fluid flow of contributions toward the joint construction of common goods, i.e., the commons. The authors define the commons as shared resources that are maintained or produced by a community or a group of stakeholders, governed according to the rules and norms of that community. The chapter elaborates on the forms of cooperation enabled by commons-centric economies, and presents a strategy for commons-based value creation and capture. It also discusses in detail the relation between commons-based economies and blockchain networks.

References

[1] Zuboff, S. (2019). "The Age of Surveillance Capitalism". Profile Books, London

[2] Malone, T. (2004). "The Future of Work. How the New Order of Business Will Shape Your Organization, Your Management Style, and Your Life". Harvard Business School Press, Boston

[3] Benkler, Y. (2006). "The Wealth of Networks. How Social Production Transforms Markets and Freedom". Yale University Press, New Haven and London.

[4] McAfee, A., and Brynjolfsson, E. (2017). "Machine, Platform, Crowd. Harnessing our Digital Future". W.W. Norton & Company, New York.

[5] Eurofound (2020). Living, working and COVID-19: First findings – April 2020, Dublin. https://www.eurofound.europa.eu/publications/report/2020/living-working-and-covid-19-first-findings-april-2020. Accessed June 7, 2020.

[6] Brynjolfsson, E., Horton, J., Ozimek, A., Rock, D., Sharma, G., and Yi Tu, H. (2020). "COVID-19 and Remote Work: An Early Look at US Data". https://john-joseph-horton.com/papers/remote_work.pdf. Accessed June 7, 2020.

[7] https://www.bitkom.org/Presse/Presseinformation/Corona-Pandemie-Arbeit-im-Homeoffice-nimmt-deutlich-zu. Accessed June 7, 2020.

[8] https://www.ibm.com/blogs/policy/facial-recognition-sunset-racial-justice-reforms/. Accessed June 7, 2020.

[9] Harari, Y. N. (2018). "21 Lessons for the 21st Century". Jonathan Cape, London.

[10] Harari, Y. N. (2020). "Yuval Noah Harari: the world after coronavirus". Financial Times, March 20. https://www.ft.com/content/19d90308-6858-11ea-a3c9-1fe6fedcca75. Accessed May 6, 2020.

[11] Arendt, H. (1957). "The Human Condition". The University of Chicago Press, Chicago & London.

2

The Future of Work in the Sixth Wave

Sofi Kurki and Markku Wilenius

University of Turku, Finland

Abstract

Our contribution takes the point of departure in the expected impact of the COVID-19 pandemic virus on the future of work and employment. We explore the future of work through two different long wave theories stating that human sociotechnical systems can be understood through long range cyclical renewal processes. While Kondratieff's wave theory addresses the long term transformation of socioeconomic systems, the approach of Malaska provides a socioevolutionary theory of social change. Both streams of work leave open the transformation of work and evolution of skills. We conclude that, given the implications of the pandemic for the nature of work and employment, there is an even greater need to focus on human skills and capabilities.

Keywords: Future of Work, Kondratieff wave theory, social change, skills, capabilities

2.1 Introduction

As we are writing this chapter in the early May 2020, there is suddenly a very different future in front of us. Due to cascading shock effect of the Coronavirus, in the United States, over 20 million jobs have been lost only in April this year, making a record level of 14,7% unemployment rate, not seen since the Great Depression. Everywhere else too, the rates of unemployment are rising. Approximately, 122 million Indians have lost their job. European governments have pumped massive amount of money to private sector in the form of direct aid or, for instance, in wage subsidy programs. Thus, 40 million jobs have been saved using those measures while the amount of debt is mounting. The service sector—which is the largest sector in most of the countries and has experienced the most rapid growth until this point—has been hit particularly hard, but it looks as if all the sectors are affected by the calamity caused by the COVID-19 virus. As global job markets are shrinking at an unprecedented pace, the real question is how this "black swan" will affect the future of work. Even if it is too early to make any conclusions, some observations can be recorded.

First, lot of measures that have been undertaken since the commencement of the pandemic, either by governments or companies and other parties, suggest that technological adaptation will move much faster than previously thought. Education sector will deploy massively new digital technologies. Companies will be increasingly using technologies that enable people to work more from home. Financial technologies (Fintech) will leap rapidly as a result of rapid increase of digital transactions and values models. Secondly, extensive travelling or transportation of goods is coming under scrutiny. There are also strong localization tendencies which may in the long run have effect on job markets.

Thirdly, and perhaps most importantly, future of work as it looks like through the lense of these disruptive events, plays a significant role in the types of skills we can call activity skills [1]. They are the kind of skills that enable people to deal with complex situations, such as the current crisis, by using more consciously our capacity to assess various future options and act upon them. We may call this specific set of skills and indeed the whole approach as futures literacy. Futures literacy comprises, first of all, our cognitive capacities to understand different development trajectories of our societies. It also involves capacity to understand how we can shift our perspective and become aware of our own assumptions about the future. Secondly, futures literacy involves our direct relationship with future. How do we create a relationship with future that is positive? How do we find our own strengths in life and project them into the future? Thirdly, futures literacy means the active capacities to be proactive toward future. We should not let future "to happen" but to be a master of our journey, even if the circumstances are rapidly changing.

Long before the current crisis, the discussion on the future of work has gained momentum due to rapid and recent advances in computation. They have led to dramatic improvements in artificial intelligence (AI) that have perhaps come as a surprise to most of the people outside of this specific field.

The development of AI and related technologies, such as robotics, are currently fueling the discussion regarding how will the societies be affected by the potential automation of jobs that have for long been considered as safe within the realm of activities that are difficult to automate, or even outsource [2, 3, 4].

Indeed, viewed on the surface it seems that the forces that have dominated much of the 20th century organizational practices such as optimizing processes, cutting costs, replacing human labor with automation, and moving production to countries that provide cheaper workforce, are gaining yet another victory by the above mentioned technological advances. They enable, for instance, yet unseen possibilities for outsourcing. Not only in terms of physical distance but also increasingly cultural divides—such as language—can be bridged by using advanced technologies [5]. Given all this, what would give us the rationale to suggest that this most recent technological development is not just the intensification of the megatrend of automation and robotization, running through the past decades, but rather the last breath of a dying paradigm?

The technological revolution we are facing coincides with what can be called as the crisis of the Western societies [6]. As Jacobs and Mazzucato, among others, have pointed out, there is a substantial amount of evidence that in the course of the past 30–40 years our global economic system has had a strong bias to increase inequality in terms of income and social benefits and make some people extremely rich [7]. According to this line of thinking, we are already experiencing the results of the impact of more traditional technologies and organizational practices that have led to widespread losses of employment in previously stable, middle-class areas. This has inspired some (e.g., Ref. [8]) to point out that it might not be the technology, but rather the problematic structures of our societies and economies whose impacts need to be re-examined. In the past, major societal changes have collided with a technological paradigm shift (the phenomenon often referred to as sociotechnical transition, see e.g., Ref. [9]). Although it has been argued that the ICT-revolution that started in the 1970s has brought with it the kind of societal dynamics that include the tendency for the economic inequality to increase, the same technology, in a more advanced and mature phase, may also contain the seeds for a very different development trajectory.

At the same time, the very foundations of human systems are being questioned by the three-fold crisis in our relationship with nature: (1) escalating in the form of global warming, (2) increasing loss of biodiversity, and (3) destruction of ecosystems. It is no coincidence that it happens right now, as we are struggling with the consequences of our technological development. It is the industrial order that has put focus on scaling up of extractive economies without any other consequential thinking that shareholder profit that has brought about the current double-edge crisis of nature and labor, and that forms the current state of societies we are observing.

In this chapter, we explore the futures of work though two long wave theories that state that human sociotechnical systems can be understood through long range cyclical renewal processes. In this chapter, we outline two basic theoretical approaches for understanding social change from this perspective and explore findings from our research on novel, human-centric organizations. These findings are analyzed as pointing to possible futures that in some regards depart radically from what is understood as mainstream organizational behavior in the present. We explore these examples especially from the perspective of skills that can be expected to be focal in the emerging wave.

2.2 The Kondratieff Wave Theory as a Model for Societal Change

One of the most influential theoretical approaches that provide a mechanism for how socioeconomic systems transform over time has been the framework of the long-wave phenomenon. It was first brought into a broader international discussion by the Russian economist Nikolai Kondratieff in the beginning of the 20th century [10]. The Kondratieff wave theory postulates that societies develop in cycles of 40–60 years, which begin by a period of growth, and end in a period of decline and depression. In Kondratieff's theory, the notion of socioeconomic cyclicality is primarily derived from data on economic activity. The wave-like behavior of economic activity can be illustrated, e.g., by the rolling 10-year yields of the Standard & Poor equity index (Fig. 2.1). From these temporal patterns we can draw out how each of the waves has been driven by specific technologies and their applications. So, we go from the birth of industrialization brought about by the invention of steam machines, to the use of steel and railroads, and further on to the electrification of our factories and homes, and new chemistry to produce, among other things, paper. From the 1930s onward, a great tide of automobiles burst onto the scene, accompanied by sharply rising consumption of petrochemicals, such as petrol for cars. The last wave, emerging in the early 1970s after the oil

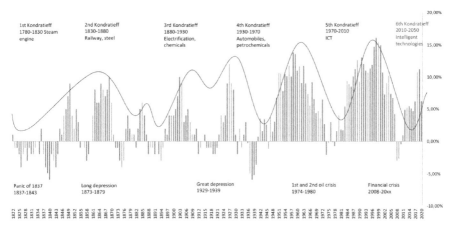

Figure 2.1   Kondratieff waves. Rolling 10-year return on the S&P 500 from January 1814 to March 2020 (in % per year). *Source*: Datastream, Bloomberg, Helsinki Capital partners (illustration), Markku Wilenius.

crisis, turned out to be the victory march for digitalization and the expansion of various communication technologies. That era came to an end with the emergence of a financial crisis that almost crashed the whole global financial system.

However, for the purpose of building on the economic pattern for understanding the social contents of the waves, we need to explain the technology uptake of each wave not only by the technologies themselves, but with some social drivers. Indeed, our argument is that the anticipatory power of the waves stems from the observation that the technology uptake that is conventionally utilized to explain the formation of a new wave can be linked with, even explained with the human interest, which by and large "tames" and directs technologies and economic behavior to a certain trajectory as illustrated in Fig. 2.2.

To understand this logic better, let us take a look at another theory that illustrates the societal change process as a series of long wave fluctuations: a theory developed by Pentti Malaska, a Finnish pioneer of futures studies [11].

2.3 Malaska's SocioEvolutionary Theory of Social Change

Pentti Malaska's socioevolutionary theory [11] is influenced by the better known Kondratieff wave theory, as well as the Maslow theory of the hierarchy of needs. Malaska's main contribution, and the main argument

THE SUCCESSION OF DEVELOPMENT WAVES
IN INDUSTRIAL SOCIETIES

K-Waves	1ˢᵗ wave	2ⁿᵈ wave	3ʳᵈ wave	4ᵗʰ wave	5ᵗʰ wave	6ᵗʰ wave
Period	1780–1830	1830–1880	1880–1930	1930–1970	1970–2010	2010–2050
Drivers	Steam Machine	Railroad Steel	Electricity Chemicals	Automobiles, Petrochemicals	Digital communication technologies	Intelligent, resource efficient technologies
Prime field of application	Clothing industry and energy	Transport, infrastructure and cities	Utilities and mass-production	Personal mobility and freight transport	Personal computers and mobile phones	Materials and energy production and distribution
Human interest	New means for decent life	Reaching out and upwards	Building maintenance	Allowing for freedom	Creating new space	Integrating human, nature and technology

Figure 2.2   Development waves in industrial societies.

in his theory, is that each new stage of (cyclical) human development, i.e., the technologies used and the social structure espoused, is guided by the need to fulfil new needs in the order of the Maslow's need hierarchy. This is important, as the psychological component introduces the notion of intentionality to macrolevel societal developments. Malaska's model makes development over time resemble a learning process where the learning goal is to achieve ever more holistic well-being. The historical stages with their specific techno-socio-economic settings are experiments aiming to reach this goal.

Figure 2.3 illustrates the key elements of Malaska's theory. The basic structure of the scheme is shared with other cyclical macroeconomic models of development, which are built around the transformational periods between more steady developmental phases. In Malaska's theory, these transformational times produce a change not only in modes of production, organizing, and economic systems but also give rise to different world views and aspirations. The key insight is that the transformation stems out of the needs created by the previous societal phase. Thus, while each of the waves has their core technologies, and their fields of applications, for the purpose of analyzing the waves, we propose these technologies also are fulfilling a human interest, which is particular to each wave. With each new wave, human

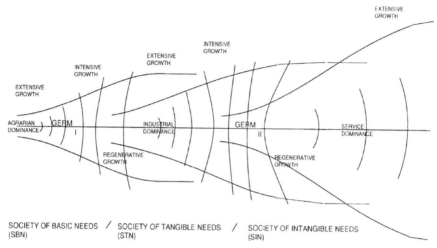

Figure 2.3 Malaska's cyclical model of societal development [11].

interests have expanded to new fields as a result of reaching a certain level of maturity in the previous wave. Again, starting from industrial revolution, the first wave helped people to reach a certain level of life quality. Clothing industry, where many of the new technologies were being applied, is a prime example of this type of development where human interest was pointed toward making it easy for people to fulfill a basic need of obtaining clothing. The second wave developed to meet very specific needs for urbanization, both by building infrastructure for the cities, as well as by enabling movement from the rural areas to the factories in the cities. The railroads allowed people and goods to be transported more efficiently than with horses. With the rise of the third wave, following the fulfillment of basic level needs, technology was used to raise the level of comfort in human societies through electrification of cities and factories. The rapid development of synthetic chemical use was also paving the way for modern life: for instance, the creation of industrial chemicals supported the households in maintaining their level of cleanness and sanitation. As we moved to the fourth wave, automobiles and petrochemicals together provided individuals technological means for mobility, thus providing certain sense of freedom.

The emergence of distributed computing systems in the wake of the fifth wave created a digital space for people to operate and communicate in. The true revolutionary force of digital technologies has empowered human beings all around the planet by allowing them to communicate on their own behalf, thus activating people to take part in different human subsystems in a very different way. This is how humans have tamed technology, originally built to solve problems in logic, to provide solutions for their fundamental human needs.

2.4 The Society of Intangible Needs

For Malaska, there have been two structurally different waves predating the one that is currently in formation. The first major sociotechnical construction was the society of basic needs, in which efforts concentrated on fulfilling the very basic needs of humans: the main concern in this era was the production of food as basis of stable human societies. The next phase was the society of tangible needs. The Kondratieff wave theory, as based on data from capitalist economies, concerns primarily this phase, as discussed in Chapter 1. The main concern of this phase was to satisfy the material needs of humanity for more comfortable living. It is the phase that is still dominating human societies at large, but from Fig. 2.3 we can see the logic in which different kinds of approaches start to emerge and build the next phase of human societies. The society for intangible needs as the next phase of the development of human societies is presented as a conceptual framework for analyzing the shift from the current paradigm to the next. Malaska uses the key concepts of *communication and technology; social vs. technical skills; organizational units;* and *progress, development, and growth* in describing the society of intangible needs [11].

Communication and technology combine to fulfill the fundamental human needs related to interaction. Communication technology is understood here in a broad sense as encompassing the basic infrastructural backbone for the society of intangible needs. Progress and development are in Malaska's scheme seen ultimately as the mental and spiritual growth of human beings. As we are moving into the society of intangible needs, the economic system from the tangible phase becomes weary and its ability to grow and prosper stagnates. Also, the perception of the benefits of constant material expansion in the physical realm has changed as a result of its cost becoming more evident. Environmental destruction becomes visible as overconsumption of basic needs becomes a key problem of the developed societies. At the social and human levels, organizational units are for Malaska a key way to conceptualize the change from tangible to intangible needs. In the next societal phase, as a result of the communication becoming the focal point of societal development, the interaction between human beings can mean, for instance, that family-like formations, not based only on blood ties, on the level of much larger entities would start forming. These family-like organizations could reclaim the traditional role as production units that were temporally lost in the society of material needs with the core family concept that strictly separated the concepts of family from societal production organizations. In the society of intangible needs, people with similar interests and values form close networks that replace the core family model of the previous phase.

Malaska did not fully elaborate in his writings the question of skills needed in the society of intangible needs. In this chapter, our goal is to explore the question of key skills for the sixth wave, in light of Malaska's vision of a society of intangible needs. To do this, first we need to zoom into the development of organizations as a context in which work is being conducted.

2.5 Evolution of Skills in Organizations

The third model that can be combined with the long wave theories for understanding the evolution of human societies, more specifically from the perspective of organizing, was originally presented by Frederick Laloux [13]. His view, echoing Malaska's framework, adopts the idea of organizational development in stages that are strongly connected with the evolution of consciousness and world view. According to Laloux, there have been four major perspectives in the history of human organizing, with the fifth one as currently emerging.

The first form of organization began to take shape perhaps some 15,000 years ago when humans began to assemble in small groups that had some practices in common (for more details, see Ref. [13]). Leadership was highly authoritarian: the group leader exercised absolute power and maintained that power by fear. Weaknesses were not to be shown. All heavy toils were performed by slaves. The cohesion of the group was based on an absolute obedience. This kind of organization mainly resembles a pack of wolves. There are still some organizations today that work in a similar fashion, such as the mafia, street gangs, and paramilitary organizations.

More stable forms of organization began to emerge with the development of agriculture, which required routine and repetition. At the same time, the church evolved as an institution that through its religious teachings maintained norms within the community. The church also brought along rigorous hierarchy. Armies proceeded to develop this model of organization still further. Also, primary and secondary schools have largely followed this form of organization.

The third form of organization evolved in the wake of modern commercial organizations. Multinational corporations evolved and an unprecedented concentration of money and power emerged on the Wall Street. Machine became the metaphor for organization, and management was guided by goal setting, KPIs, and optimization. Organizations are strictly meritocratic and constantly seeking new markets and innovations. This form of organization is still prevalent today.

A few decades ago, a new kind of organization began to appear whose key value lies in information and knowledge. These organizations are very much about building up a "we" spirit, and for the most part they are nonprofits. Further, 10 years ago, it was estimated that these organizations were worth more than 1 trillion dollars, making them the world's eighth biggest economy[1]. For these organizations, the role and meaning of culture takes precedence over everything else.

It is only during the 21st century that a new, fifth type of organization has begun to emerge that clearly stands apart from all the previous models. These organizations are based on three key distinctive principles:

(1) *They are self-directed and minimize hierarchy.* They recognize the need for management, but not so much managers. This model is most effective in complex environments, which is increasingly true in our societies today.

(2) *They work on the premise of a broad and comprehensive image of the human being.* Employees are expected to bring in all their capacities to the organization and its work. The role of emotions in communication is recognized and appreciated. Organizations encourage their employees to pursue their own goals. The development of the organization is firmly based on feedback systems.

(3) The third distinctive feature is that *organizations are managed with a very sensitive ear.* Traditional strategic thinking is less important, and the organization works much like a shoal of fish, keenly aware and sensitive to the environment and one's own emotional sensations. These organizations often have a far-reaching mission, but the way in which this mission is pursued can vary widely.

The key aspect of Laloux' analysis is that all the five models of organization are still in active use today. Of the five, the latest, fifth phase

---

[1]https://www.globalpolicy.org/component/content/article/176/31937.html

of organizational development links with ease with Malaska's society of intangible needs. The image of a society of intangible needs illustrates a society where the basic assumptions that have driven the development of the industrial phase have changed to include a more nuanced understanding of human capabilities, values, and even virtues. Here, especially communication and related social skills are seen as playing a key role. Ultimately, the shift is in the mindset that includes the way the essential qualities of humanity are understood. The change has been visible already for some time in global value surveys, and it is increasingly affecting organizational and societal practices in the form of self-management, or citizen initiatives in fields as separate as science, art, and urban planning [6]. What is common to all these indicators is the questioning of authorities, increasing reliance on social and peer networks, and increasing focus on self-expression as a motivational factor. The development of AI is also giving impetus to philosophical discussion around the nature of humanity, its future, and moral responsibility between human beings displaying different intelligence levels [14]. It may therefore be argued that the emphasis on the social dimension of human intelligence is fundamentally a product of the advancing technology: the fundamental effect of AI is perhaps not the threat it poses to rote jobs, but rather the possibility of evolving understanding it offers for humanity about itself. If being a human is no longer distinguished by the superior intelligence among other creatures, what will be the alternative for human status?

2.6 Key Features of New Work

In the previous phase of organizations, the structure of the working life was based on a notion of a career, built on professional degrees obtained through formal education. The posts were assumed to be more or less permanent, and only the demand and salary levels were expected to be subject to change (usually to increase) as a factor of time, as this was understood to reflect experience. In the recent years, this model has been broken by unstable working markets, increasingly volatile business environments, and policies encouraging flexibility and change. The result, in the experience of many, has been increasing feelings of insecurity, and only for some the changes have brought about feelings of opportunity and dynamism [15].

However, the notion of a traditional career path has also been challenged by increasing demands directed at work as a provider of meaningful life.

Values related to self-expression are on the rise globally as described by Ronald Inglehart's world value survey, based on longitudinal data[2]. Meaningful work is a hot topic in organizational research as well, providing insights on what kinds of elements provide experiences of meaningfulness in the workplace. Robust findings attach meaningfulness to such attributes of work as autonomy, benevolence, (utilizing and developing of) competence, and feeling of togetherness [16, 17, 18, 19]. Although the stability of the career might not be in contrast with these values, many other aspects of traditional, industrial era workplaces, often characterized by stiff hierarchies, little change, and difficulties in perceiving the impact of one task from the perspective of the whole process, might well be.

The optimal life, both at work and outside of it, for many would be to have the means available to explore new possibilities and learn while maintaining the psychological safety provided by stable social networks and belonging to a community. This assumption is backed up by research commissioned by the World Economic Forum that shows the most important motivating factors in the digital economy to be *creativity, learning new things, and being able to have impact on others* [20]. These interests are interestingly reflected on the diminishing value placed on traditional rewards, such as monetary compensation[3].

As a conclusion from the above, one may argue that factors affecting the future of work are both endogenous and exogenous, i.e., Malaska's notion of a self-produced demand for change is supported by technological development that functions as an enabler for many of the things that such new paradigm of work requires.

2.7 Essential Work Skills in the Society for Intangible Needs

Basic skills that are required from any individual in an organization can roughly be divided into two categories: (1) *analytical skills*, including knowledge of the substance of the work, related technical capabilities, and a deep understanding of the principles governing the field that together amount to expertise, and (2) *social skills* that enable an organization to function as a

---

[2]http://www.worldvaluessurvey.org/wvs.jsp

[3]Although here the context of research as enquiry into digital behavior may have an effect, it cannot be directly transferable to the motivations related to workplace behavior.

collective body of different individuals. The organizational development over the 20th century has gradually led to a great emphasis on the first category, where development and utilization of the individual substance expertise have been seen as the most valuable asset, both for individuals and their organizations. Social skills were taken as granted as something essentially human, and organizational structures were assumed to remove the need for much individual social involvement in organizational dynamics[4].

However, the higher expertise the work demands, the more the value is placed on independence, creativity, and personal involvement. These elements often clash with hierarchical protocols and structures. Therefore, there has been increasing pressure toward forming organizations that are more adapted to the change in the workforce, and in the nature of the work toward highly specialized knowledge work. In knowledge work, the end product is often a result of a creative collaboration between several experts. This setting poses completely different requirements to the ability to take part in and benefit from social processes than the more mechanical tasks in typical industrial organizations. Thus, while the main attention still remains predominantly on individual expertise, in many organizations there has been a noticeable attempt to modify practices so that more efficient knowledge work could be possible. The most radical form of this development is the so-called self-managed model of an organization. Today especially the IT-sector has embraced the idea, but it is increasingly penetrating other fields as well. In fact, the model has proven more efficient than the traditional hierarchical structure even in very basic industrial work, as demonstrated for instance by a tomato-producing factory Morningstar in California that engages in fully self-organized principles in its operations. This implies that knowledge work as a category is too narrow to describe the kinds of organizations that feel the pressure to relax the traditional hierarchical structures. However, it is not only the organizational practices, but crucially also individual skills that need to be developed in the new context.

What are then the skills that have been underappreciated in the industrial paradigm, but that are expected to be crucial in the next phase? Typically, the role and responsibility of the individual are at the center of attention for any low- to nonhierarchical organization. This responsibility, however, can only be realized in a social context. Taking responsibility of one's work requires

---

[4]This assumption, of course, was proven wrong already at the very beginning of enquiry into organizational behavior, with the famous Hawthorne studies showing various effects that are based on the social aspect of work in contrast to the clearly defined instructions the work that was based on.

an ability to negotiate the role within the greater group of other individuals, argue for one's own views and debate about them in a constructive manner, and finally to take into account others' views enough to be able to formulate a common understanding of the goals the process should aim for.

Already in Malaska's theory of society of intangible needs, the anticipation of the specific skills that will be essential in the new paradigm, is an important element in building up the vision. His argument, at the time, was that a trend toward tasks requiring human interaction skills, at the cost of technical skills, is indicative of the direction the whole society is developing. Elsewhere, a more recent argument has been brought forward that in the robot age, meaning a society where the majority of today's tasks can be automated by highly sophisticated AI, the skills required by an employer will be the so-called metaskills. Neumeier labels these under five categories: (1) feeling, (2) seeing, (3) dreaming, (4) making, and (5) learning [21]. In part, some of the authors advocating emphasis on metaskills rely on the argument that as the future at this very special point of sociotechnical evolution makes anticipating the needs of the future job market especially challenging, the focus in education should be on providing students with skills that help them to adapt better to any potential scenario.

Counter to these visions, some have proposed exactly the opposite, making a clear statement that technological prowess is the key to success in tomorrow's job market [20, 22]. A safe bet, it seems, might be assuming a position in the middle ground: a law-like historical trend over the human societies seems to be developing ever more detailed specialization of expertise, and assuming the complexity of the societies to increase further, we can expect this to continue. However, the same trend of increasing complexity also requires more and more of what can be called contextual understanding. This means that abilities traditionally attributed to only the very top management now need to be shared by all the employees of an organization. These include understanding the "big picture," the context where the organization is situated and the challenges are faced, the trends affecting the future of the organization, and how they can be taken advantage of for achieving the organization's goals.

Contextual skills are important for each individual because increasingly, the strength of an organization lies in its ability to practice ambidexterity: developing and innovating while maintaining the basic business [23, 24]. This is much more challenging if the requirement to understand the operational environment is placed only on the chosen few at the top of the organization. In contrast, when these issues are understood at the level of individual employees, distributing power and responsibility about the development of the organization's activities becomes a way to manage complexity. Considering the toughening competition within most fields, and the tendency to automate rote jobs, the value added by individual experts will more and more come from this kind of an anticipatory mode of work. Therefore, essential combination of social skills presented above and these so-called contextual skills is key for maintaining a functional organization, while the *technical skills* still provide the backbone for the substance an organization is engaging in.

In terms of individual skills, we are suggesting that as we are moving from the era where the skills needed were those that cater for basic needs— as identified by Maslow—to the next level of needs as requested by the complexity of our world, we are deploying more and more skills that can be called *active skills* [1]. Based on understanding the way a society of intangible needs functions, we may identify the following core skills and capabilities that are needed as essential metaskills for the life and for the job markets of the future:

First, we are referring to what can be called *complexity skills*. These skills essentially deal with managing the overwhelming amount of information, while keeping our thinking clear and consistent. Here the reference is clear to our ever digitalizing and globalizing environment that delivers growing amount of information that should be able to make sense. As societies and each of us, the individual environment is becoming more complex, we need ways to improve our resilience to cope with fast changing circumstances with flexibility, capacity to manage uncertainty, and ability to act in complex decision-making situations. As the world grows ever more interconnected and develops more technology, we have more information than ever to deal with. This tests our capacity to keep our minds clear and focused and not get led astray with too many impulses and stimuli. Our sensemaking ability becomes critical, and we can only cope with the current reality if we learn how to think in systems i.e., find connections and cause–effect relationships between individual phenomena. Developing our complexity skills helps us understand the structures of society and discover opportunities to influence matters within that complexity.

Secondly, we seem to be in growing need for *creativity skills*. This is to question how do we enhance our capacity for finding new and unconventional solutions to the problems we face. Moreover, we need to be more creative to deal with everyday life in the job markets. The old idea of staying your life with one of two employers is dying as job markets are becoming more volatile. According to the US Bureau of Labor Statistics, the average amount of jobs people undertake in the course of their career in the United States has grown to 12 [25]. From these statistics, we can deduce that elderly people tend to have longer terms of jobs than younger generations.

Creativity ultimately denotes to human quality and as such is an important resource for us to use, as machines, robots, and AI hold the potential to release us from monotonous and repetitive tasks. Creativity is a unique human feature that technology or AI cannot replace. Thus, creativity skills aim at strengthening our creative and critical thinking faculties and draw on our inner capacity to use our intuition. Creativity does not necessarily require great flashes of genius—it is more akin to the ability to find new, sometimes surprising solutions to existing problems. Moreover, as the famous creativity researcher argued, creativity unites cognitive processes with environmental factors and individual properties of a single human being [26].

Thirdly, as globalization and connectivity intensify in our world, we need increasingly *empathy skills*. We need to have broad understanding and know-how of other cultures and their specific ways of seeing the world. Being empathetic means also having capacity to collaborate. It seems as if cultural competence become a crucial asset for individuals and organizations to be successful in the interconnected world [27]. The question ultimately here is, how do we grow our capacity to see issues from the perspective of others. Using and developing empathy skills mean to focus directly on our capacity to reach out and loosen our innate bias to be self-centered in our thoughts and actions. It has been argued that the positive leaps forward in our civilization came about due to our capacity for emphatic behavior [28]. Even the current COVID-19 crisis has promoted a great amount of empathy as people in our societies become more aware how interconnected they are.

Fourthly, we need to understand that a lot of what will drive our societies moving forward is dealing with issues that help us become more sustainable. Thus, we need more *planetary living skills*. The question is how do we create a healthy relationship with our physical environment? A lot of new businesses and jobs will be created on those areas that can solve problems that our past activities have created. To build these new activities, we need to understand and experience how nature actually works and affects us (e.g., how plants grow and what they offer us). In today's world where people are increasingly raised in urban environments, it is important to maintain a living relationship with the biosphere surrounding us. As we have moved into the Anthropocene era, we need to realize that humanity's impact on the planet has become so significant that we have to take responsibility to secure the future of our planet.

When we move our attention to organizations, we start to observe those behaviors emanating that result from collectively utilizing the above-mentioned skills and putting them into action. The skills become principles and practices that are consistent with future-orientated mindset. All in all, key features that illustrate both the skillset required in tomorrow's organizations, and the organizational principles, can be condensed into seven points that we have called it the CARE model [29]:

1. *The level of quality in ways to share information and collaborate.* Great emphasis needs to be placed into making sure all the necessary information about the functioning of the system is available for everyone in the system. This means not only opening up information that has traditionally been reserved for a limited group of mostly senior executives, but also enabling seamless peer-to-peer communication through new channels enabled by social media. Also, the information flows from the clients to the providers need to be unrestricted. This emphasis on information, from a systems perspective, signals an understanding of the importance of the feedback loop for the self-correcting ability in organization's functions.

2. *Cultural fit and social skills becoming important in recruitment.* Even if the technical skills and competencies relevant to the tasks performed by the organizations evidently constitute key criteria for recruitment, increasingly, "cultural fit" and social skills are in even a more decisive role for maintaining the social organization functional. When these issues are the more important, the more self-managed the work itself becomes: when most of the responsibility resides within the independent individuals, it becomes very important that the individual characteristics of employees are suitable for their task.

3. *Continuous and inclusive focus on innovation.* Innovation is not compartmentalized to a specific department in the organization, but rather it is a feature of everybody's work. This is reflected in approaches that emphasize addressing the specific needs of every client as opposed to offering ready-made solutions to fit all. Also, the purpose of the organization is driving all organizational activities. The organization is geared toward all talent and know-how being directly applied for the benefit of the clients instead of utilizing it unnecessarily to bureaucratic

functions that can be solved in other ways. This point is a reflection of systems thinking being applied in the organizations design. Learning in an organization can only take place if we allow human-centric principles, instead of hierarchy and bureaucracy-based principles to be the mode of operation. This cannot take place unless people are given full freedom and responsibility in the ways they organize their work.

4. *Passion for radical solutions.* Novel organizations are committed to nonstandard solutions. They are not seeking the quickest and most convenient solution to everything but rather want to ensure that all decisions are made consciously, with reflection. This is particularly relevant because in the sixth wave, the world simply needs more radical solutions. We have entered a new socioeconomic era that seeks ways to radically reinvent the central pieces of our system, be it the question of social services, counteraction to environmental destruction, or mending our corrupted political systems. It is much more probable that those radical solutions come from organizations that encourage the innovation to take place at every level.

5. *Commitment to long-term thinking.* This has been expressed, for instance, by stating that the goal of an organization is to be a flourishing company after 200 years. Even more importantly, this thinking has been brought to a very practical level. The demise of many organizations today is that they keep doing things that they know are wrong. In the organizations we have studied, however, people and teams are encouraged not only to innovate but to "fail fast." If something is not working, it is not continued. People and teams are constantly encouraged to challenge their own practices. This is the way to increase the sensitivity to changes in the operating environment.

6. *The ability to take risks.* This is all about creating a culture where it is safe to take risks of all kinds: bring up issues that are uncomfortable, and "go and experiment with something that has not tried before." People need to feel safe to be able to use their full capacity. A strongly risk aversive culture leads to sticking to old rules and habits. Of course, risk taking needs to be coupled with high levels of ethics. One of the key factors bringing down the productivity of organizations is the level of fear. In a fear-based culture, people are not encouraged to bring in new initiatives. In the sixth wave organizations, taking and allowing risk is considered a conscious act. From a systems perspective, risk taking allows an organization to question its current practices and to open up to new opportunities.

7. *A focus on futures.* Currently, the predominant way to assess performance is focused on past merits. Organizations are thus led into the future by measuring the past. In modern human-centric organizations, there is a strong belief that future needs to be allowed to happen. Two main guidelines are applied for this approach. Firstly, strong principles (rather than targets) need to be established which the organization will adhere in all of its behavior. Secondly, different initiatives for the organization's future are set to be tested among coworkers. If an idea proves successful among the people, then the whole company might adopt the new idea. In such a way, the future actually emerges all the time in various forms. In a fundamental way, for these organizations, the future always remains open. It is neither fixed into the present nor is it tied to a singular vision about the future: nobody in the organization "owns" the future. From a systems perspective, allowing the future to unfold means that a sufficient amount of input is allowed at all times to enable multiple futures the possibility to occur.

2.8 Conclusion

As mentioned earlier, our world is currently going through a tough period of transformation set forth by the dramatic implication of global pandemic caused by the COVID-19 virus. This transition will accelerate a lot of those developments where new technologies are applied to save resources, be they money, time, or nerves. As a result of the accelerated pace of change, not only lot of jobs will become obsolete but also new jobs will be created. Individuals, organizations, and countries will face challenges that push them to focus on those skills and capacities that enable them to be competitive. Our essential conclusion is that to stay competitive, we need to focus on particularly those skills that make us better as human beings. We need to focus more on human skills.

In this chapter, we have explored findings from novel human-centric organizations in the frameworks of Kondratieff wave theory and Pentti Malaska's theory of human needs guiding societal transformations. The findings from human-centric organizations are compatible with the general technological trend of replacing rote work with automation, and a human need for finding balance, stability, and opportunities to learn. An overall motive in human-centric organizations is enabling an environment where an organization can truly be a reflection of the capabilities of its members. Work serves the function of what is ultimately a fundamental human need to be able to contribute to one's community. Thus, we argue, work itself will find novel fields of operations in new services while simultaneously automation and robotization will continue to eat low value-added manufacturing work. As the sixth Kondratieff wave will proceed toward the middle of this century, the transformation we are observing in working life will find ever new manifestations in our societies.

Three aspects sum up the key implications of the society of intangible needs for the capabilities required by organizations in the sixth wave, as a systemic shift from competition to collaboration reorganizes the working life: Firstly, expanding the realm of creativity: how do we organize our working as to enhance creativity and being more responsive to our client's needs? It means more autonomous teamwork, and more heterogeneous organizations. Secondly, how do we place learning as a key target for our organizations? This means more subtle means of communication and less information hoarding, a typical bias to industrial age companies. Thirdly, how do we ensure the possibility for anyone in the organization to have an impact on the issues that concern her/him? It means more direct communication and less hierarchy.

If all these key questions become sufficiently addressed by the cultivation of those virtues that we have proposed as seminal properties of organizations, then we definitely are on the right path. As we have suggested in this chapter, there is an inherent logic regarding how a society moves from one stage to the next. The future of work is just one aspect of this larger change dynamics of our societies. We need to observe how our ways of work are changing, understand what kind of implications this change has, and take action that is necessary to stay tuned to the demands of our time.

In spite of all dystopia about the end of the work as robots and automatization swallows our jobs, there will be work in the future. For many organizations and individuals, it is just going to be different in terms of how it is done and with what motivation and aspiration. But the biggest change that will happen in the long run, we believe, is why we are doing work. Obviously, we need to do that for living, but what else?

As we are moving toward the society of intangible needs, we shall find how the whole society will gear toward values that respect our rising awareness of how we should run our organizations. The future of work will include a dramatic downgrade of hierarchies and communication blocks that were often quintessential part of the industrial era. The future of work will also include the rise of stakeholder considerations, away from self-centric way of running business. Above all, it will incorporate more human-centric approach to run organizations, understanding what aspire humans of this century. As the future will prove, that is the only smart way forward.

References

[1] Wilenius, M. and Pouru, L. (2020). "Developing futures literacy as a tool to navigate an uncertain world". In "Humanist Futures: Perspectives from UNESCO Chairs and UNITWIN Networks on the Futures of Education". Paris: UNESCO.

[2] Brynjolfsson, E. and McAfee, A. (2011). "Race against the Machine: How the Digital Revolution is Accelerating Innovation, Driving Productivity, and Irreversibly Transforming Employment and the Economy". Digital Frontier Press.

[3] Frey, C. B. and Osborne, M. A. (2013). "The future of employment: How susceptible are jobs to computerisation?" OMS Working Papers, September 18. https://www.oxfordmartin.ox.ac.uk/downloads/academic/future-of-employment.pdf

[4] Webb, A. (2019). "The Big Nine: How the Tech Titans and Their Thinking Machines Could Warp Humanity". New York: Hachette Book Group Inc.

[5] Baldwin, R. (2019). "The Globotics Upheaval: Globalization, Robotics, and the Future of Work". New York: Oxford University Press.

[6] Wilenius, M. (2017). "Patterns of the Future: Understanding the Next Wave of Global Change". London: World Scientific.

[7] Jabobs, M. and Mazzucato, M. (2016). "Rethinking capitalism: An introduction". In "Rethinking Capitalism: Economics and Policy for Sustainable and Inclusive Growth", Jabobs, M. and Mazzucato, M., (eds.), Hoboken, New Jersey, NJ: Wiley/The Political Quarterly Publishing Co. Ltd.

[8] Mazzucato, M. (2018). "The Value of Everything: Making and Taking in the Global Economy". London: Penguin.

[9] Grin, J., Rotmans, J. and Schott, J. (2010). "Transitions to Sustainable Development: New Directions in the Study of Long-Term Transformative Change". Routledge Studies in Sustainability Transitions.

[10] Kondratieff, N.D. and Stolper, W. F. (1935). "The Long Waves in Economic Life". The Review of Economics and Statistics, 17(6), 105–115.

[11] Malaska, P. (1999). "A Conceptual Framework for the Autopoietic Transformation of Societies". FUTU publication 5(99). Turku School of Economics and Business Administration, Finland Futures Research Centre.

[12] Malaska, P. (1998). "Sociocybernetic transients of work in the late-industrial period. USA and Finland as the empirical cases". SA XIV World Congress of Sociology, Montreal.

[13] Laloux, F. (2014). "Reinventing Organisations. A Guide to Creating Organisations". Brussels: Nelson Parker.

[14] Harari, Y. N. (2017). "Homo Deus: A Brief History of Tomorrow". London: Vintage.

[15] Pugh, A. (ed.) (2017). "Beyond the Cubicle: Job Insecurity, Intimacy, and the Flexible Self". Oxford: Oxford University Press.

[16] Martela, F. and Riekki, T. J. J. (2018). "Autonomy, competence, relatedness, and beneficence". Frontiers in Psychology. DOI:10.3389/fpsyg.2018.01157

[17] Martela, F. and Ryan, R. M. (2016). "The benefits of benevolence: basic psychological needs, beneficence, and the enhancement of well-being". Journal of Personality, 84, 750–764. DOI: 10.1111/jopy.12215

[18] Deci, E. L., Olafsen, A. H. and Ryan, R. M. (2017). "Self-determination theory in work organizations: the state of a science". Annual Review of Organizational Psychology and Organizational Behavior, 4, 19–43. DOI: 10.1146/annurev-orgpsych-032516-113108

[19] Deci, E. L. and Ryan, R. M. (2000). "The "what" and "why" of goal pursuits: human needs and the self-determination of behavior". Psychological Inquiry, 11, 227–268. DOI: 10.1207/ S15327965PLI1104_01

[20] World Economic Forum (2016). "The Future of Jobs. Employment, Skills and Workforce Strategy for the Fourth Industrial Revolution". Global Challenge Insight Report. http://www3.weforum.org/docs/ WEF_Future_of_Jobs.pdf

[21] Neumeier, M. (2013). "Metaskills. Five Talents for the Robotic Age". San Francisco: New Riders.

[22] Hakovirta, M. and Lucia, L. (2019). "Informal STEM education will accelerate the bioeconomy". Nature Biotechnology, 37(1), 103–104. DOI: 10.1038/nbt.4331

[23] O'Reilly, C. and Tushman, M. (2016). "Lead and Disrupt. How to Solve Innovator's Dilemma". Stanford: Stanford Business Books.

[24] March, J. G. (1991). "Exploration and exploitation in organizational learning". Organization Science, 2, 71–87.

[25] US Bureau of Labor Statistic: Number of Jobs, Labor Market Experience, and Earnings Growth: Results from a National Longitudinal Survey. https://www.bls.gov/news.release/pdf/nlsoy.pdf. [Accessed 10 May 2020].

[26] Csikszentmihàlyi, M., (1990). "The domain of creativity". In "Theories of Creativity", Runco, M. and Albert, R. (eds.). Sage focus editions, Vol. 115. Thousand Oaks, CA: Sage Publications, pp. 190–212.

[27] Wilenius, M. (2006). "Cultural competence in the business world: a Finnish perspective". Journal of Business Strategy, 27(4), 43–49. DOI: 10.1108/02756660610677119

[28] Rifkin, J. (2010). "The Emphatic Civilization: The Race to Global Consciousness in a World in Crisis". New York: Jeremy P. Tarcher Inc.

[29] Wilenius, M. and Kurki, S. (2017). "K-Waves, reflexive foresight, and the future of anticipation in the next socioeconomic cycle". In "Handbook of Anticipation. Theoretical and Applied Aspects of the Use of Future in Decision Making", Roberto Poli (ed.). Cham, Switzerland: Springer Nature.

3

Emerging Technologies and Working Life

Risto Linturi[1] and Osmo Kuusi[2]

[1]R. Linturi Plc, Finland
[2]University of Turku, Turku, Finland

Abstract

In this chapter, we describe a framework for technology foresight and anticipation of a technology driven societal transformation. We also present current findings of a research program which started in 2012 and is still ongoing. The study divides anticipated not-incremental technological development into 100 separate categories and societal and human goals into 20 main categories linking these two kinds of categories. A societal goal category, i.e., global value producing network (GVN), has an impact on the evaluated importance of a technological category, i.e., anticipated radical technology (ART) that is relevant to the societal goal. On the other hand, the relevant technology category has an impact on the content of the societal goal category (GVN) described as a transformative sociotechnical regime. The method unites demand pull and technology push and solves many problems inherent in previous foresight methods. As a major result, we can show that all 20 GVNs may face a paradigm-level transformation due to anticipated progress of enabling ARTs and goal-seeking behavior of actors in each GVN. This will challenge existing professions, the way we organize work, our tools and capabilities, and even the need to work for money. The purpose of this tool is to anticipate these changes. Time perspective of this study is the next 20 years.

Keywords: Foresight, anticipation, generic technology, transformation, sociotechnical regime, technological breakthrough, weak signal

3.1 Introduction

The research has been commissioned by the Parliament of Finland's Committee for the Future. Its aim has been to anticipate societal and technological transformations to enhance understanding of the need to change existing regulations, infrastructural and educational goals. National universities of applied sciences and vocational schools have been major users of the results when directing their own research and education.

We start by describing the radical technology inquirer (RTI) method and continue with an overview and the results. As the research covers all technologies and all organized functions of the society, the main emphasis here is only on the major points. Most of the described changes impact on work by changing work environment, organizational goals, destroying professions, and creating new ones. More details are provided in Ref. [1].

Foresight studies often concentrate on specific business areas or potential consequences of selected developments. This has been the typical approach of technology foresight Delphi studies. In these studies, experts have evaluated hundreds of statements concerning particular future uses of technologies. Technology foresight Delphi studies have followed three basic principles: (1) touching upon future topics in the form of statements, (2) having a type of assessment or questions concerning these future topics, and (3) feedback of previously given answers by other respondents be it in rounds or roundless [2]. This is, however, not good practice, if the main focus of the study is generic technologies as in the RTI. According to Kokshagina et al. [3] generic technologies empower the design of several technological applications for multiple markets, create value across several domains, and contribute to the disruption of existing industries.

RTI method is inclined to cover all major technological breakthroughs and based on them try to anticipate especially important generic technologies in 20 years perspective (Fig. 3.1). This can naturally succeed only partially as via no process can any researcher fully become aware of all potential innovations due to, i.e., trade secrets or human bounded rationality. The societal potential of technology-enabled innovations can still be evaluated through all major goals of socioeconomic actors. This interaction is also inevitably incomplete due to similar reasons but avoids many of the problems inherent in more narrow approaches.

RTI analyzes the interaction between technology development and societal structures via application ideas. This impacts both on the credibility of technological developments and anticipated transformations. Potential

Radical Technology Inquirer (RTI)

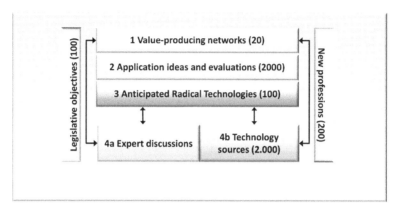

Figure 3.1 The framework of radical technology inquirer method [1, page 67].

sectoral transformations provide the basis for anticipating new professions and new regulative requirements. Due to the wide-ranging realm of the study, in our data collection, we have relied on continuous crowd sourcing, expert panel discussions, and continuous assessment of research and development sources described and listed in detail in the latest full report [1].

3.2 Technological Development and Societal Goals

3.2.1 Anticipated Radical Technologies

When one aims to gather anticipatory information on all fields of technologies, the main challenge is not knowing what one does not know. There are no search terms and no experts you could rely on to find such information you have no questions for. The method selected in this study was the continuous crowd sourcing. Continuous process is beneficial as the novel breakthroughs can be recognized from science and technology news streams and attentive and interested specialists, and technology enthusiasts can easily be motivated to participate both in evaluating the new findings and contributing their own findings from their own areas of interest and their own favored sources. At the moment of writing this chapter, there are c. 3.000 people participating in a Facebook group set up for this purpose in 2013. They provide daily posts containing links to scientific publications and news of advancement in radical breakthrough technologies. Continuous monitoring allows picking up new themes when they are published. The variety of sources, viewpoints, interests, and expert backgrounds produce a good foundation for further discussions and evaluations. Additional people are invited by any group member when a member believes that some area is not amply followed. All postings are moderated by the authors of this chapter.

The latest published report [1] contains 1600 selected links to radical technology breakthrough news. These findings are grouped into 100 categories and each category is described as an anticipated radical technology (ART). Each ART solves some technological or socioeconomic bottleneck and thus enables new innovations or further market diffusion of existing innovations. Examples of ARTs are driverless vehicles, solar fuel, urban farming, material scanner, easy GMO-technologies, standards for modular robotics, or efficient and light batteries.

ART does not describe any existing technology as such or even a novel or expected incremental innovation unless it enables radically new potential for applications. ART describes anticipated technology through its capability, which may arrive in market within 10 years and have major impact within 20 years. Each ART is evaluated based on the source material. Then the maturity score is given based on available material and the description of the ART. The maturity development is followed based on consecutive reports. In addition, the maturity score is also used to assess the probability of ART impacts.

The identified radical technologies are categorized into seven levels of maturity; each level increasing the successful solution to the ART and potential market breakthrough. The seven maturity levels are the following:

1. The scientific principle that enables the breakthrough in the area of an application has been proven to be possible at the level of a theoretical model, published by a research team from a reliable research institute. The observation is recent, and there appears to be academic interest in the subject.
2. The scientific model enabling the breakthrough in the area of an application has been verified with concrete laboratory testing, and the operating principle has been demonstrated in a peer-reviewed publication. The research groups are funded, and the academic motivation is clear. Progress appears to be taking place.
3. The laboratory prototype leading to the breakthrough in the area of application has been demonstrated. The laboratory prototype clearly

indicates that some material problem that previously prevented the breakthrough has been solved and the necessary functionality has been achieved. Possible problems relating to production costs or durability have yet to be solved at this level. The research groups are funded, and the academic or commercial motivation is clear. The progress is recent, and the new opportunities presented are significant.

4. Proof of concept (PoC). A functional prototype has been scientifically or commercially demonstrated to meet the requirements of commercialization in terms of its functionality and viability for production. In terms of its functionality, the prototype must clearly exceed the benefits of previous solutions or involve lower production costs than previous solutions after the economies of scale have been taken into account. The progress is recent and clear.

5. A competition situation with several well capitalized market actors developing the production prototype after the PoC phase, or a production prototype clearly in the finishing phase, with investments made to launch production. The progress is fast-paced.

6. Products are being delivered to customers in increasing amounts. The economies of scale are expected to reduce prices considerably. The areas of application are expected to expand. R&D activity is at least partly based on internal financing, and the expansion of production has been clearly financed. The progress is recent.

7. Competing products are being delivered, and the customer demand is on the rise. Competition is internally financed to a significant degree, and the growth is financed by major industrial companies, a wide clientele, or venture capitalists. The development of product properties involves a significant known potential.

The collection of 100 ARTs is divided into 10 groups as shown in Table 3.1. This grouping is mainly intended to help users to find suitable categories and illustrates here the areas included and is a practical but not essential part of the method.

Initially, all findings were grouped loosely into these 10 categories and then in an iterative manner by authors further into 100 ARTs. The formulation of each ART aims to target the main bottleneck, which maybe solved according to the findings. Some of the ARTs are completely novel, however, i.e., solar panels are not a new invention. They are included because there is an evidence of a major development potential based on the radically new production methods with radical consequences in several sectors.

Table 3.1   Groups of technological breakthroughs.

Technology groups	
1	Instrumentation and telecommunication
2	Artificial intelligence and algorithmic deduction
3	Digitalization of sensory data and processing
4	Transport, mobility, and logistics
5	Production of products and services
6	Material technology
7	Biotechnology and pharmacology
8	Energy technology
9	Digital crowdsourcing platforms
10	Globalizing technology interfaces

This potential includes radical price decrease and also major weight, flexibility, and efficiency gains that open totally new applications and restructuring of various business practices and working life.

3.2.2 Global Value-Producing Networks

The technology push is a poor viewpoint by itself. It is evident throughout history that new knowledge usually appearing in the form of novel technologies and social innovations enables change. However, equally important as it is to know what is possible, is to understand the versatile goals that people have. Technology or knowledge by itself has no impact unless at least some people wish to use it. Thus, a motivational analysis is a crucial part of the technology foresight, and it needs to consider both existing and anticipated motivations for all relevant actors in each area.

Many of the socioeconomic structures in the society have been formed around core technologies. We can talk about dominant regimes where value is produced using certain typical methods and tools [4]. Each regime has its own typical business ecosystems, a regulative framework, and core values. These have been called sociotechnical regimes.

As new technologies arise, we can imagine more efficient or otherwise better ways to fulfill the same goals. These are called challenger regimes, and change from the dominant regime toward the challenger regime is called the regime transformation. Challenger regime definitions used in the research are founded on the enabling ARTs and supported by the collected source material. They have arisen in crowdsourcing discussions, stakeholder workshops, and seminars during the years 2013–2018 as the most plausible contestants replacing the dominant regime. Basic requirement for a challenger regime is that the source material in the form of ARTs must support anticipation of the alternate sociotechnical regime as both plausible and interesting to the necessary actors. In most cases, early signs of a transformation can already be seen.

The potential transformations are not necessarily self-evident from the base data. Their final forms have been evaluated in stakeholder workshops arranged by several Finnish ministries and major innovation system organizations separately for each GVN. However, as is the case in all anticipation concerning human systems, the extent of a transformation contains unpredictable elements and should be considered as partial directions at best.

For the purpose of this foresight study, society has been divided into 20 GVNs, each with an easily understood basic goal. The passenger transport system is a good example. The basic goal is to move people from one place to another. The dominant regime is private cars and the public mass transport operated by drivers. As a challenger regime, we have selected the autonomous transport as a service.

Table 3.2 shows the 20 GVNs, which are used in the study as a motivational framework and context of the socioeconomic transformation including the existing dominant regime and the proposed challenger regime. Jointly the 20 challenger regimes represent a major and unprecedented transformation potential throughout the whole society. They are organized in such a manner that each GVN has a clear and understandable prime goal and existing structures and actors, but also a potential radically new way to satisfy the same goal. These transformations have major impact on how our work is organized and performed, but also on the skills required.

Besides the basic goal, which is satisfied either by the existing dominant regime or the challenging regime, each GVN description includes moderating values, opportunities, and threats of the challenger regime and required enablers for the challenger regimes' success. The necessary maturation of novel technologies (ARTs), required regulative changes, and examples of new processes and professions are also described. New professions illustrate the organizational and educational challenges caused by the transformation. The time frame when these transitions are expected to have considerable impact is within the next 20 years. Challenger regimes are expected to become important but not necessarily dominant by then.

Table 3.2   Twenty global value-producing networks (GVNs).

GVNs with their most potential transformations

	GVNs	Dominant regimes	Challenger regimes
1	**Passenger transport**	Private cars operated by a driver, public transport	Autonomous transport as a service
2	**Logistics**	Transport operated by a driver, repetitive automated loading	Autonomous transport, smart loading robotics
3	**Manufacturing of goods**	Industrial, centralized, repetitive manufacturing	Robotized, decentralized, discrete manufacturing
4	**Sustenance**	Agriculture, food industry, distribution channels	Urban cultivation, robotic local food preparation
5	**Energy supply**	Centralized and fossil energy sources, peaking power plants	Renewable, decentralized energy sources and energy storage
6	**Materials**	Mining-based products, energy-heavy process industry	The circular economy, renewable materials
7	**Built environment**	Traditional construction and maintenance	Robotized construction and maintenance
8	**Exchange**	Brands, physical retail locations, hierarchies, B2B2C	The reputation economy, e-commerce, P2P, C2B2C
9	**Remote impact**	Telephone, television, Internet, social media	VR/AR, avatars, and other remote controls
10	**Automation of work**	Centralized automation and human-steered machines	Decentralized robotics based on AI and crowdsourcing
11	**Work and income**	Salaried employment related to specialization and exchange	Cooperation, self-sufficiency, microentrepreneurs
12	**Health care**	Health-care system, general health recommendations	Self-diagnostics, gamification, individual nutrition
13	**Redressing disabilities**	Institutional, outpatient and family care, cheap assistive devices	Robotics, AI, avatars, artificial organs, crowdsourcing
14	**Acquiring information**	Certified research, reports, news	AI, crowdsourcing, personal instruments and applications
15	**Proficiency and its proof**	Educational institutions and qualifications, on-the-job learning	Flipped learning and independent learning, AI, proficiency demonstrations

(Continued)

Table 3.2 Continued

GVNs with their most potential transformations

	GVNs	Dominant regimes	Challenger regimes
16	**Producing experiences**	Focus on producers and consumers, mass entertainment, tourism	Games, shared VR, AR, interaction, AI
17	**Safety and security**	Material safety in society, social security	Decentralized, individual and crowdsourced safety and security
18	**Collaboration and trust**	Guaranteed by authorities, brands and hierarchies	Peer-to-peer trust through platforms and transparency
19	**Existential meaning**	Work, position, social network	Achievements, likes, participation, communities
20	**Power structures**	Regional power structure, opaque power	Subject-matter subsidiarity, location independence

Next, the proposed transition in each GVN is characterized.

1. **Passenger transportation:** The goal is to transport people. The challenger regime is the autonomous transport as a service. The robotization of transport frees the driver. Modes of public transport can be reduced in size, departures can be made more frequent, and individual mobility can be offered as a service (MaaS) at an advantageous price and taking the challenge of climate change mitigation into account. The use of shared resources will become simpler than before when you can call a mode of transport to come when you need it and leave it to continue its way to the next person when you no longer have any need for it. This also makes it possible to use an increasing variety of modes of transport to implement a travel chain. A good example of this is the rapidly expanding shared use of bicycles in cities. Autonomous transport also enables the easy and affordable individual mobility of people without a driver's license.

2. **Logistics:** The goal is to transport goods and materials. The challenger regime is the autonomous transport and smart loading with robotics. Existing freight transport is mainly based on repeated automation and human labor. Smart robotization enables cost-efficient sorting and autonomous transport of goods. This allows individual items to be transported cost-effectively from the manufacturer to the consumer and for material flows to be delivered to manufacturers as needed. The goal is to achieve MaaS in logistics, with an open digital platform linked to

freight transport. It consists of the identification of goods, crowdsourced participation, and a high degree of robotization in the delivery of goods from the manufacturer to the place of use and from there to recycling thereby taking the challenges of sustainability into account.

3. **Manufacturing of goods:** The goal is availability of goods. The challenger regime is the robotized, decentralized, and discrete manufacturing. With the advancement of robotization, a production structure that is based on the present-day mass production and the hierarchic distribution may become decentralized so that an increasing proportion of goods will be manufactured near customers in accordance with their individual needs. Flexible production lines and contract manufacturing are already a step in this direction. Digital manufacturing combines design and manufacturing tasks and increases the subcontractor's responsibility through a model-based definition, making it a part of the seamless production entity. Production and service are drawing closer to each other.

4. **Sustenance:** The goal is human and pet sustenance. The challenger regime is urban cultivation with robotic food preparation. The challenger regime is based on aquaculture that is primarily carried out under artificial lighting in a closed space in cities and factories as well as the insect husbandry, biotechnical food production and robotized, decentralized and customized food orders that takes the challenge of a climate change mitigation and resilience into account. By replacing the cultivation of fields and animal husbandry with urban farming, production can be made continuous, local, and need-based, and food production will no longer require a major industrial intermediate phase. Indoor farming and new GMO techniques will make genetically manipulated food safe, efficient, nutritious, and tasty.

5. **Energy supply:** The goal is sustainable energy especially taking the challenge of the climate change mitigation and need for locally independent energy supply into account. The challenger regime is renewable, decentralized energy sources and energy storage. New energy sources and storage techniques are increasingly challenging previous energy solutions. The utilization of the solar energy in local production of electricity and heat is becoming increasingly inexpensive at the global level. The costs of the solar energy are estimated to eventually fall below the production costs of competing forms of energy in most areas. Converting solar and wind power into gaseous and liquid fuels promises to solve the energy storage problem in a way that will

simultaneously make part of the electricity grid and district heating network unnecessary.

6. **Materials:** The goal is access to materials. The challenger regime is the circular economy and renewable materials motivated much by the challenge of the climate change mitigation. The current dominant regime primarily relies on the mining industry, forestry, or other energy-intensive process industry and mass production that use large raw and limited material streams. The challenger regime offers materials and nanotechnology-based metamaterial structures that are renewable and efficient from a process technology or functional perspective. We are also heading toward the circular economy, low-temperature biological processes, advanced structures, and a need-based economy of smallness.

7. **Built environment:** The goal is availability of suitable facilities. The challenger regime is robotized construction with novel functionalities and a robotized maintenance. Construction has been very people-intensive, but robotization makes it possible to automate casting, brickwork, painting, the transfer of materials and many installations and finishing tasks. However, 3D printing of buildings, bricklaying robots, and painting robots are in the pilot phase. Construction is becoming automated, both in the prefabrication carried out at factories and in on-site construction. Maintenance tasks are becoming robotized, starting from lawn mowing and cleaning of floor surfaces. Automation will progress to the maintenance of streets and roads, in addition to more demanding tasks. In addition, the needs for built environments will change greatly due to major changes in people's everyday routines, technological and business practices including local production, logistics, and virtual reality.

8. **Exchange:** The goal is transfer of rights. The challenger regime is the reputation economy, e-commerce, crowd sourcing, and peer collaboration. Traders are becoming global and crossing national borders. E-commerce delivers the goods to the customer through a general logistics network. Online shops serve as their own marketing channels, and they do not use external channel marketing in the manner of traditional trade. Brand trust is replaced by customers' product-specific likes and recommendations based on comparisons on social media. Due to decentralized logistics and high level of automation in e-commerce, the subcontractor networks can be massively distributed. Payment traffic operates electronically, and the means of payment may

extend to cryptocurrency. Exchange between companies is becoming an activity between machines that is guided by data systems and artificial intelligence. The decisions and emotions of end users and other profiling may affect this activity as much as internal data of component suppliers and companies. Similarly, to consumer trade, trade between companies is also transitioning to platforms.

9. **Remote impact:** The goal is remote influence without the need to travel. The challenging regime is the virtual and artificial reality, robotized avatars, and other remote controls. There have been great changes to the means of remote influencing. Global IT platforms allow individuals and organizations to be disturbed or controlled by means of communication. Viruses, social media, cryptocurrencies, and the dark web are the growing channels of influence. Remote-controlled robots and missiles, the disturbance of devices connected to the Internet, and artificially created infectious diseases are the means that have previously been possible to major national operators, but they are becoming accessible to increasingly small operators. Active participation and physical labor are also becoming possible on an everyday level by means of telepresence.

10. **Automation of work:** The goal is reducing cost of work. The challenger regime is decentralized robotics based on AI and crowd sourcing. Human labor is being replaced by machines at an accelerating pace. Compared to today's repetitive automation or specialized machine power requiring continuous human control, new machines have increasingly versatile capabilities while still being autonomous. Robots can move independently from place to place and can, e.g., process materials, take measurements, or guide humans according to a need or wish detected by the robot itself. Robots and 3D printers manufacture a wide range of items. Mobile robots can take care of logistics or perform environmental maintenance tasks, such as taking cuttings from grapevines, weeding, or ploughing snow. Machines can also make observations, model things with the help of artificial intelligence and learn to recognize deviations and the best operating models. A focus in the regulation of the automation of work will be on the challenge of climate change mitigation.

11. **Work and income:** The goal is securing one's own well-being. The challenging regime is communal cooperation, self-sufficiency, and microentrepreneurship. The boundaries of paid work, entrepreneurship, and subsistence economy will change. Artificial intelligence, the sharing economy, and robotization make the important tools and expertise

required for productive work easily available. Differences in work productivity and need for specialization may decrease as a result. The platform economy also facilitates occasional employment relationships, crowdsourcing, and sharing of the results of communal work. Smaller and smaller organizations are capable of doing an increasing number of things. The economy of smallness is making work tasks more comprehensive and transferring them from the sphere of paid work to a subsistence economy. In the future, consumers will be able to use their own and shared smart machines to make objects, clothes and even medicine, buildings, or vehicles for themselves. This signifies a return to a family-centric subsistence economy.

12. **Health care:** The goal is long and healthy life. The challenger regime is self-diagnostics, gamification of health, and individual nutrition. In the near future, people will be able to use household equipment to independently monitor the condition of their bodies more closely than the laboratories of central hospitals are now able to do. AI, assisted by an online expert, will be able to provide a statement on any necessary lifestyle changes and need for treatment. The means of proactive care and early recognition of symptoms will improve, particularly in the self-care. DNA sequencing is becoming so inexpensive that, in addition to our genome, it will likely become practical in the near future to also determine the state of our protein metabolism and the body's bacterial strain as well as the genome of the cells we eat as food on an individual basis. The research into extending the human life span is at the cusp of a potential breakthrough. The concretization of this research may lead to a radical increase in voluntary treatments.

13. **Redressing disabilities:** The goal is to optimize one's functional ability. The challenger regime is individual assistance with robotics, AI, avatars, and crowd sourcing. The visually impaired have access to machine vision applications that allow them to recognize objects, spaces, routes, and facial expressions. The hearing impaired can utilize speech recognition. AI interpreting speech as a text or visualized sounds by other means. Prosthetic legs and arms, even eye prostheses are becoming robotized, and they can be connected to the human nervous system or even to the brain with a brain implant, allowing senses and movements to be controlled with thoughts. Wearable robotic legs that strengthen the body's own movements are available for people with muscle weakness or partial paralysis. An artificial pancreas produces the amount of insulin required by the body in real time. A sleeve placed around a weak heart

to strengthen its beating is under development, as are artificial muscles, subcutaneous measurement devices, and medicine dispensers. Children, elderly people, and other people with a limited functional capacity may have a personal AI assistant that monitors their actions, warning them of dangers, and reminding them of past and future events. This type of assistant can monitor traffic and the rest of the person's surroundings, the function of the body, and the actions of other people.

14. **Information acquisition:** The goal is to have reliable information. The challenger regime is the personal access to AI, crowdsourcing, instrumentation, and expert applications. The world is becoming immediate and seemingly straightforward. Studies are published immediately as draft versions before a peer review. Messages are often even purposefully misleading. Search engines and AIs help us evaluate the truthfulness of the messages we receive. Search engines provide us with information about our local region. With personal measurement devices, we can study the things around us and compare the results with other people interested in the same phenomena through cooperation networks. Materials and substance compositions can be identified with optic means. DNA readers will become household items, and many other optical, biomechanical, and electronic devices will make it very easy for us to identify molecular compositions in our food, environment, and even the air exhaled by other people, for example.

15. **Proficiency and its proof:** The goal is proficiency and its recognition. The challenger regime is flipped and independent learning and proficiency demonstrations utilizing various digital technologies. In addition to making learning materials widely available, information technology also allows teaching situations to be experienced regardless of time and location. It is increasingly possible to transit from teaching to learning by trial and error. Simulators allow students to test their own skills at a level of difficulty that contributes to their learning. AI can correct their performance and provide an adequate number of stimuli to motivate them to learn. As automatic test environments, AIs and simulators can also help demonstrate an adequate proficiency. It is apparent that the proficiency will be gathered from an increasing number of sources in the future, and that it will also become outdated much more quickly. Instead of degrees, various other demonstrations, such as certificates related to private microdegrees, peer evaluations, customer evaluations, and demonstrations of skill and competitions have already become important methods of evaluating proficiency in many fields.

Management of reputation and reputation networks, as well as search for results and reputation assessments generated by search engines and AI, are increasingly important. Journeyman's demonstrations, portfolios, and mentoring are an essential part of this perspective.

16. **Producing experiences:** The goal is access to meaningful experiences. The challenger regime is the interactive utilization of computer games, shared virtual reality, and AI. Technological advancement has freed up time for experiences. Many experiences have also become time- and location independent. People are already listening to music while walking down busy streets, and soon music videos will be shown as holograms around the walker. Digitalization is making experiences democratic. Computer games are an example of a new way to participate in stories. A game provides a shared environment and stimuli for shared experiences. New technologies provide ample opportunities for the adventure travel, art experiences, and nature trips as highly realistic experiences. VR glasses can be used to create an illusion of an actual presence in a virtual world. Experiences can be very powerful, even deceiving the body. Producing physical experiences is becoming increasingly easy with robots. In the future, robots will be able to prepare gourmet meals, paint art on the walls, massage our shoulders, or play the violin. Sex robots have become a phenomenon of their own in Japan, for example. AI makes it possible to recognize human emotions and preferences. With this information, AI is able to adapt its operation to each individual's way of thinking and produce the desired emotions and experiences.

17. **Safety and security:** The goal is regularity of living environments. The challenging regime is decentralized, individual and crowdsourced safety and security. The centralized monitoring that targets producers and protects consumers no longer functions well with society making everyone into a producer and with borders becoming porous and virtual. An increasing part of safety and security must be produced closer to the individual either by robots and AI or crowd sourcing. In addition to our data, our identity can also be stolen. People are increasingly efficiently threatened and blackmailed from beyond national borders by strangers. As our everyday objects become smart, also threats may become physical. Cars drive themselves; robot cooks prepare meals. These functions maybe hijacked. Cyber threats may involve manipulation of our own technical appliances or remote-controlled devices that transport toxins or bombs to our vicinity.

18. **Collaboration and trust:** The goal is cooperativeness and synergy of people. The challenger regime is peer-to-peer trust trough platforms and transparency. Most of our current transactions are built on hierarchical trust and cultural norms. This will not be adequate due to the changed nature of the modern collaboration. If participation is effortless or in one's own interest, there is no need to fear free riders. When the stakes required for collaboration are uneven, actions are required to increase trust. The opportunity for revenge is offered to the party who invests first as a potential sanction that it can impose. The simplest sanction is a public reprimand. On a more general level, we can refer to the reputation economy. The parties provide their reputation as collateral and increase their reputation based on successes. Transparency of the activity increases trust. Transparency can also be increased by certifying each product, space, and service with a unique identity and making the actions taken in relation to this certification available to other parties in a cloud service maintained by a trusted third party. A blockchain is the most notable new method of transparency.

19. **Existential meaning:** The goal is a meaningful life. The challenger regime is based on achievements, social media likes, and communal participation. Grand narratives have changed, self-realization and serving others gained importance. Digitalization and robotization unlock new types of comprehensive work roles in which meanings are the focus, and the work is performed for the members of one's own community either in data networks or the local community. Gamification, when carried out correctly, can increase the meaningfulness of existence and actions at the individual level. The Internet has materially facilitated contact with people who are meaningful to us. Doing things together and achieving things in digital communities will gain ground as the possibilities offered by the virtual world expand. The virtual world will grow into a meeting place and work environment for increasingly large groups. With AR glasses, we can be constantly in contact with our immediate circle as though we were in the same place. This may once again shift our set of values toward a more collective emphasis. At the same time, this may reduce the importance of external status and status achieved with money. The scope of sharing has grown to a global level, making it easy to find like-minded people. Besides keeping a pet, people will find a new significance in raising an AI, building virtual worlds, having a hero's reputation in a game world, and helping those in need in games.

20. **Power structures:** The goal is optimization of joint decisions. The challenger regime is a subject-matter subsidiarity and location independent power structures. Digitalization frees us from being tied to a time and location and automates simple decision-making. The demarcation between public and private tasks should be revised in the changed situation. Participatory decision-making is a step back toward this subject-based proximity. Each subject-specific administration could be independent of geography, and competing administrations could exist for all tasks that are not tied to a location. A great number of administrative matters can be handled routinely. Executing administrative regulations with the help of AI, gamified systems, simulation, and crowdsourcing would make regional administrations significantly better equipped to perform their own duties efficiently and perfectly. Platform cooperatives in data networks are bodies managed by their members for the organization of joint administration and marketing. In platform companies and platform cooperatives, the decision on the acceptance of each assignment is made by a member. Platform cooperatives could be provided with considerably large responsibility for many of the administrative tasks related to present-day democracy.

3.2.3 Anticipating the Importance of Each ART and Likelihood of the Challenging Regimes

As the plausibility of GVN transformations relies on the enabling ARTs, it is essential both to evaluate which ARTs are necessary in each challenging regime, for which purposes the ARTs are being developed, how they are funded and progressing, and how credible it is that they mature for the purposes in each proposed transformation.

Each ART is evaluated against each GVN goal. Application ideas for technologies related to ARTs are crowdsourced as a part of the study. Ideas have been collected from tens of thousands of online comments and added with insights from dozens of seminars and workshops where the early forms of the GVNs and ARTs have been presented. Numerous ideas are directly based on the crowdsourced 1600 source articles describing radical technology breakthroughs.

If there is an idea or some other weak signal that the anticipated radical technology could somehow aid in achieving the goal of a value producing network or its transformation, a respective score is awarded to the ART.

The rules are designed in a manner to promote generic ARTs over those, whose benefits are limited to few GVNs only. A simple stepwise logarithmic scoring was selected and tested to give high scores if applied in initial phase to known major drivers. Internet in 1993 was one of the test cases. Scoring Internet with these rules with the knowledge one had in 1993 would have resulted in high enough total over all GVNs to gain considerable attention. A study based on an early version of this method [5] showed more generic and better scoring ARTs to mature faster than those with lesser scores and they also have greater transformative power due to providing novel interaction between different GVNs.

Each GVN gives a score to each ART according to the following rule:

- 1 point is assigned if the development of the technology basket potentially delivers concrete benefits that make it worthwhile to apply the technology to this use.
- 3 points are assigned if the development of the technology basket potentially delivers material benefits with regard to a value related to the value-producing network's goal or is part of a whole that materially promotes the actual goal. In this context, a material benefit would be an economic impact at the level of €2–20 per person yearly or, on an individual level, an average impact of 1–10 person-hours on everyday life per year.
- 5 points are assigned if the development of the technology basket potentially delivers significant benefits with regard to a value related to the value-producing network's goal or is an important part of a whole that promotes the actual goal in a transformative way. In this context, a significant benefit would be an economic impact at the level of €20–200 per person yearly or, on an individual level, an average impact of 10–100 person-hours on everyday life per year.
- 10 points are assigned if the development of the technology basket potentially results in a transformative impact, within the meaning of the description of the value-producing network, on the way the value-producing network's goal is realized. On an annual level, the potential impact must exceed €200 per person yearly or impact weekly the everyday life of over 10% of the population.
- 20 points are assigned if the development of the technology basket is potentially a necessary part of the most important transformative impact on the value-producing network's operating model. On an annual level,

the potential impact must exceed €200 per person or impact weekly the everyday life of over 10% of the population.

Total score for the potential impact of an ART is calculated as a sum of the scores from each VPN. The probable impact or the final score is calculated multiplying this potential score with maturity of the ART. This means we reduce the plausibility of a success from those ARTs, which are not yet mature enough to enter the market.

Top 25 ARTs scored against GVNs are shown in Table 3.3. Few things can be noticed easily from Table 3.3. All top technologies have a potential impact on several GVNs, which is natural as this was the goal of the scoring system. But several of them are also essential or very important enablers in many GVNs. This creates a dynamic interaction between GVNs, which is routed via ARTs. The GVNs benefiting from an ART's early application boost that ART's development and makes it more accessible even to another GVNs applications. This means that a reliable anticipatory method cannot study a single GVN without considering all GVNs to get a good assessment of that single GVN.

Table 3.3 The scored values of anticipated radical technologies.

Anticipater Radical Technology ART	Maturity	Passanger transport	Logistics	Manufacturing of goods	Sustenance	Energy supply	Production materials	Built environment	Exchange	Remote impact	Automation of work	Work and income	Healthcare	Redressing disabilities	Aquiring information	Proficiency & it's proof	Producing experiences	Safety and security	Collaboration and trust	Existential meaning	Power structures	Total potential value
Neural networks and deep learning	5	10	10	5	10	3	3	5	10	5	20	20	10	5	20	10	10	5	5	5	20	955
Autonomous cars and trucks	5	20	20	0	3	5	0	10	10	5	10	5	0	10	5	0	5	20	3	3	0	670
Environment scanning & positioning	7	20	20	3	3	0	0	3	0	5	10	3	0	3	10	3	5	5	0	0	0	651
AI performing local work on global basis	4	10	10	5	5	0	3	3	10	20	5	5	10	3	5	20	5	5	10	5	20	636
DNA reading and writing (full genome)	7	1	3	0	20	0	10	0	3	3	0	0	10	5	20	0	0	10	0	5	0	630
Rapid development of photovoltaics	7	5	5	5	10	5	20	5	10	1	3	3	3	0	0	3	0	5	3	3	0	588
Commercial platforms for sharing economy	6	10	5	0	5	0	0	5	10	20	3	10	0	3	3	5	3	0	5	5	5	582
Speech recognition/synthesis and interpreting	6	3	3	0	3	0	0	3	10	5	5	10	3	10	5	5	5	5	10	5	3	558
Real time 3D-modelling of environment	6	20	20	10	5	0	0	5	0	1	5	0	0	3	5	0	5	5	0	3	3	540
Material scanner - hyperspectral camera	5	5	10	5	10	0	5	5	5	5	5	3	10	3	10	1	0	10	5	1	5	515
Transportable batteries and supercondensators	7	20	10	0	0	5	0	5	0	3	5	0	0	5	5	0	10	3	0	0	0	497
3D-printing of things	7	1	5	20	0	0	3	3	3	3	5	5	0	1	0	5	3	3	3	3	1	469
VR-glasses, MR-glasses and virtual reality	6	3	3	5	0	0	0	5	3	5	3	10	1	3	3	5	10	1	3	10	5	468
Radical growth in computing powed	4	10	10	5	5	0	5	5	10	5	10	5	5	0	5	0	10	5	5	0	10	440
Quadcopters and other flying drones	7	1	20	3	3	0	1	5	1	3	5	3	0	0	5	0	1	10	0	0	0	427
Personal health diagnostics systems	5	5	3	0	10	0	0	5	1	3	3	0	20	10	5	3	3	5	3	5	0	420
Verbot/chatbot, talking/corresponding robots	5	5	5	0	3	0	0	3	5	10	10	5	3	0	5	5	10	5	3	3	3	415
M2M trade and other online commerce	5	5	3	5	5	1	5	20	5	5	3	0	3	1	0	0	5	5	0	0	5	405
Cloud computing and storage services	7	3	5	5	5	1	0	0	10	3	3	3	0	0	3	0	3	3	5	0	5	399
Ubique environment and internet of things	4	3	10	10	5	0	3	10	10	5	5	3	3	0	10	0	5	5	5	3	3	392
LED-farming, robotic farming	6	0	5		20	5	3	5	3	3	3	3	3	0	0	3	3	0	5	0	0	384
Facial / emotion recognition and projection	4	3	1	0	3	0	0	3	5	10	10	3	3	3	5	10	10	5	5	10	5	376
Smart glasses, AR-glasses and augmented reality	5	0	5	5	10	0	0	10	3	5	0	5	0	3	3	5	10	5	3	3	0	375
Pattern recognition and other AI platforms	5	3	5	5	5	0	0	5	10	5	10	5	0	0	3	3	3	3	5	0	3	365
Global wireless broadband networks	5	5	5	0	0	0	0	10	3	5	5	3	0	1	5	5	5	5	5	5	5	360

3.3 Conclusions

When considering the ARTs and GVNs, we clearly see that the transformations within GVNs are better suited for considering impacts to working life and ARTs are just technical parts of the dynamics. Work is mostly not about using one tool but participating in a process where various tools are important. This is what GVNs describe—transformation in the ways we produce added value. ARTs do empower these transformations and thus to anticipate the changes we must gain insight to the plausibility of the ARTs too. A large number of plausible new professions (c. 200) and changed skillsets were outlined in Ref. [1].

A defining characteristic of a GVN is the prime goal. As one can notice, the goals are formulated so that they are subjective. They reflect basic needs in such a way that we can estimate them as relatively constant for the duration of two decades within the developed world. Both shared societal goals and qualitative goals are included in the framework as moderating values. Regulative suggestions are added when individual goal-seeking actions under existing regulations would create societal problems or threats to other people and also when transformation would be hindered by existing regulations.

This kind of an approach concentrating on individual goals leads to a realistic foresight assessment at least within the time frame of two decades unless some truly unforeseen developments have extraordinarily rapid and major societal impact. This cannot be ruled out but such major impacts can be considered improbable. Even the SARS-CoV-2 pandemic which is ongoing at the time of writing this chapter seems to favor the challenging regimes rather than stop their advancement. An example direction being favored is digitalization.

The method can be used to anticipate both problems and positive outcomes. Both can create new professions and change working life. For this anticipation to work in detailed level and produce plausible results, we must be open to all changes. We must, e.g., award scores to ARTs not based on their usefulness to the society as a whole but rather based on their impact whether it is positive or negative. If the nuclear bombs were invented now, they would gain high scores even if we all wished they never had been invented.

High scores must in this method be interpreted as a sign of potential change. We should focus on the potential and consider how the negative impacts can be mitigated and the positive impacts enhanced. And, naturally, we must understand that some people, areas, and organizations may be harmed even while most others would benefit. Solving these contradictions belongs to political debate while this tool only aims to raise awareness. And when considering working life, we must understand how numerous professions such as the police, fire fighters, soldiers, burglars, and locksmiths cannot be understood unless considering the contradictions in various interests. Technology development creates new contradictions and also new professions, which can only be understood through these contradictions.

Looking at the scoring of the ARTs one can notice the strong logarithmic scale. We aimed for a scoring method, which would have given high scores to Internet in 1993. As such, Internet could not have been seen as imperative in any single GVN. But easily one could have imagined useful applications in each GVN. Thus, Internet could easily have risen at least amongst top 25. Today, we can see the high impact the Internet has on the working life.

This method of scoring favors generic innovations. We have shown in Technological Change 2013–2016 [5] as a preliminary result how the genericity may partially explain the speed of the technology maturation. The funding for R&D may come from different independent sources and development may pass on to another realm when it is ripe enough to appear as a low hanging fruit there.

The main reason for this kind of scoring is that generic technologies cause major changes in the sociotechnical regimes. They tend to change horizontal structures to lateral ones, crossing and breaking sectoral borders uniting their previously separate platforms. This requires the attention of regulative authorities, strategic thinkers, and investors. This also requires attention of educators and those planning their studies and careers.

As we consider the 20 GVNs, they are nearly similar in 2018 report [1] compared to the original 2013 version [6]. It is, however, very clear that the challenging regime is now defined much more clearly including arising new professions, service structures, required regulations, business opportunities, threats, and values. The most important point is that the challenging regime has become much more plausible due to the required technologies maturing quickly.

Almost all 20 challenging regimes are under-regulated and at best, only arising. Very few educators are teaching the skills that are required in the potential professions of these challenging regimes. When we think about a society where this kind of fundamental change takes place, it does create tensions. Skills are underdeveloped and change can be opposed. Regulations maybe used to hinder change and favor old players' values. This creates conflicts. Even when the new structures emerge and start luring capital away from the dominant regime, total production tends to drop as the old production capacity is not maintained and the new cannot yet meet demand.

This kind of major paradigm changes are usually referred to as the Kondratieff cycles. The old cycle ends with lowering interest rates followed by major and long political, economic, and even military crisis. It is difficult to see, how we can avoid these phenomena if even a part of the regime transformations described here gain major traction during the next decades.

References

[1] Linturi, R., Kuusi, O. (2018). "Societal Transformation 2018–2037: 100 Anticipated Radical Technologies, 20 Regimes, Case Finland". Publication of the Committee for the Future 10/2018, Parliament of Finland.

[2] Aengenheyster, S., Cuhls, K., Gerhold, L., Heiskanen-Schüttler, L.M., Huck, J., Muszynska, M. (2017). "Real-time Delphi in practice—A comparative analysis of existing software-based tools". Technological Forecasting and Social Change, Volume 118, 15–27, May.

[3] Kokshagina, O., Gillier, T., Cogez, P., Le Masson, P., Weil, B. (2017). "Using innovation contests to promote the development of generic technologies". Technological Forecasting and Social Change, Volume 114, 152–164, Jan.

[4] Schot, J., Geels, F. (2017). "Strategic niche management and sustainable innovation journeys: Theory, findings, research agenda, and policy". Technology Analysis & Strategic Management, Volume 20, 537–554. DOI: 10.1080/09537320802292651

[5] Linturi, R. (2016). "Technological Change 2013–2016: Preliminary Investigation of the Development of Radical Technologies After the 2013 Review". Publication of the Committee for the Future, Parliament of Finland, 2/2016.

[6] Linturi, R. et al. (2013). "Suomen sata uutta mahdollisuutta: Radikaalit teknologiset ratkaisut". Publication of the Committee for the Future 6/2013, Parliament of Finland.

4

On Humans, Artificial Intelligence, and Oracles

Emilio Mordini

Responsible Technology SAS, Paris (France), and Health and Risk Communication Center, University of Haifa, Haifa (Israel)

Abstract

Our epoch is fascinated by human-like minds and by the fantasy of intelligent machines, which might surpass and substitute human intelligence. Computational machines were initially invented for unburdening humans from tedious and unrewarding tasks, often doing repetitive tasks better and quicker than bored human. Today, things are changing, and the mounting trend is to produce intelligent machines imitating human skills, such as intuition, emotions, sentiments, capacity for perceiving atmosphere and context, capacity for sense of humor, and so. The potentiality of next-generation artificial intelligence (AI) is thus in the limelight of the scholarly discussion and raises public concerns. In this chapter, I will develop a new argument, based on Floridi's *Diaphoric Definition of Data*, to demonstrate that computational machines might replicate human intuitive skills but they cannot exactly duplicate them. The chapter does not aim to grade natural vs. artificial intelligence, not yet to raise any ethical consideration on AI, rather it only aims to show the inherent limits of AI and its applications. Finally, I will conclude that the current debate about the hypothetical risk that AI might one day surpass human intelligence is largely misplaced.

Keywords: Intelligent machines, AI, GOFAI, next-generation AI, esprit de géométrie and esprit de finesse, explosive logic, paraconsistent logic, intuitive skills, unthinkability

4.1 Introduction

Almost no day goes by, one does not read articles in magazines and newspapers about terrific progress achieved by research and applications on artificial intelligence (AI). All industrial sectors, as well as finance and services, are investing into AI, *"ABI Research forecasts that the total installed base of devices with Artificial Intelligence will grow from 2.694 billion in 2019 to 4.471 billion in 2024 (...) There are billions of petabytes of data flowing through AI devices every day and the volume will only increase in the years to come"* [1]. Tangible and economic reasons are not, however, the sole justification for such a fashion. AI is in the Zeitgeist. As other historical periods were fascinated by robots and automata (think of the 18th and 19th centuries, and the flourishing of interest in artificial human-like creatures, from the Mechanical Turk chess player to the Golem), our epoch is fascinated by human-like minds. There would be even an evolutionary substrate to justify this interest. Since 2005, American technologist and futurist, Ray Kurzweil [2], has argued that AI is entering a "runaway reaction" of self-improvement cycles, increasingly producing more and more intelligent generations of devices, which will end up by surpassing all human intelligence. Finally—suggests Kurzweil—AI is the next step of evolution, destined to take over humanity. More recently, Swedish philosopher Nick Bostrom has suggested that, sooner or later, an artificial super intelligence could replace the human species [3].

This fantasy of substitution can be traced back in the mass culture no less than to 1968 Kubrick's "2001: A Space Odyssey" and A. Clarke's corresponding novel with HAL 9000, the fictional AI, playing the role of the villain. The plot is well known, HAL 9000 tries to take control of the spaceship to fulfill its mission to Mars, although this implies to kill human astronauts; the story unfolds against a background hinting at a vague, spiritual, epochal shift. Indeed, the fascination for AI is inseparably connected with fantasies about its evil, or at least dangerous, nature. Why do we fantasize so much of a machine intelligence going amok and taking power on humans? Could these fantasies ever become a reality? Could AI ever take over human minds and surpass human intelligence? In this chapter, I will try to provide an original answer to these questions.

4.2 The Sorrento Counterfeiters

Let's start with a real story about machine intelligence. This story begins—or rather, it ends—on May 16, 2017, when the Italian police took down a ring of money counterfeiters in Sorrento, near to Naples. This event does not seem to be that remarkable, if only because counterfeiting is not rare in the Neapolitan area. There are historical reasons for that. In the 19th century, Naples was an important hub for small, very sophisticated, publishing houses, skilled in offset printing, lithography, specialized in art books, and limited artist editions. As the business became increasingly controlled by major international publishers, Naples artisanal industry died. Many skilled printers found themselves unemployed and reinvented themselves in the counterfeit market. They gave birth to families of counterfeiters, handing down the knowledge for perfect counterfeiting from father to son. Still today, some of the best counterfeiters in the world live in Naples. For instance, in 2006 a Neapolitan forgery ring put on the market in Germany a considerable amount of counterfeit euro banknotes, perfectly imitated except they were 300 euros banknotes, a nonexistent denomination. However, these banknotes passed as real and circulated for some months.

The Sorrento ring was a small ring, made up of a bunch of teenagers and a few senior skilled counterfeiters, who recreated 10 euros notes, with perfect, artistic, precision. In small groups, made up of one adult and some youngsters, they went in the most frequented stretches of Neapolitan coast. The group pretended to be a class trip; the adult simulated to be a teacher and the youngsters his students. They searched for currency exchange machines. The fake teacher gave some counterfeit banknotes to the fake students, who changed them. In such a way, they succeeded in changing between 1,000 and 2,000 euros per day. They were all arrested with charges of forgery of money, spending, and introduction of counterfeit money. This story would be quite trivial except for one thing: they could not be prosecuted. Each note bore a visible printed caption "specimen," and they were printed scrupulously respecting Italian legal rules concerning sample banknote reproduction. Legally speaking, those people were not counterfeiters[1]; it was not their fault if machines are stupid.

---

[1] At the end, the court convicted the senior counterfeiters of illegal possession of currency grade printing paper, which is a minor criminal offense, while the teenagers went free.

Currency exchange machines are "stupid," they have almost nothing to do with AI; engineers can undoubtedly explain why they were not able to "see" the caption "specimen" and how this bug can be fixed, but this is irrelevant to my argument. Think of the tremendous artisanal talent with which the banknotes were counterfeited, the brilliant and straightforward stratagem conceived to escape the law, confronted with the limited amount of the fraud itself. Think of the lovely pantomime that counterfeiters and teenagers played. Why? Pretending being a school trip was redundant; they did not need such a trick to change banknotes, it could draw even too much attention on them. It was not a rational choice; it was pure love for staging, probably the same love for staging which drove the other forgery ring in 2006 to invent 300-euro banknotes. You need to look at the big picture to understand: the beauty of Sorrento and the history of counterfeiting in Naples are all part of this amusing story. Machines can detect fake banknotes, but over and above technical bugs, will they be ever able to see the big picture? Can machines understand the poetic nuances of a Neapolitan crib of the 17th century? Or can they grasp the art of the *Bamboccianti* (puppet makers), 17th century genre painters of everyday life in Rome and Naples? In their landscapes, any single detail is carefully depicted with a cloying perfection, but what truly matters is the whole, the details are misleading. What counts in the story of the Sorrento ring—as well as in the Neapolitan crib and the *Bamboccianti's* paintings—is the totality; only by understanding the totality, one can catch the meaning of what is facing. Could a machine ever understand this? Mechanical devices lack the capacity for perceiving the atmosphere, feeling the context, appreciating the situation in a single glance, which is instead so essential to human beings[2].

4.3 The Good, the Bad, and the Ugly

Let's consider another example: this time, it is an invented story. The story is based on a classic 1966 Spaghetti Western film directed by Sergio Leone, *"The Good, the Bad and the Ugly."* The Blondie (The Good) (Clint Eastwood)

---

[2]A further example of machines' inability to perceive totality is provided by computational humor, a branch of AI, which aims to produce a computer model of sense of humor. Based on a natural language generator programs, AI can invent jokes and puns (not that funny, admittedly) and recognize when a human being jokes, but it cannot "feel" the comical, because AI lacks two essential qualities to appreciate it, the perception of the whole and the sense of timing (comic time) [4].

is a professional gunslinger; Angel Eyes (The Bad) (Lee Van Cleef) is a hitman; and Tuco (The Ugly) (Eli Wallach) is a wanted outlaw. They all know that there is a stash of gold buried in a cemetery. Tuco only finds out the name of the cemetery, while Blondie finds out the name on the grave. Angel Eyes knows they know the location of the gold. Finally, the Good, the Bad, and the Ugly find themselves in the courtyard of the cemetery where the gold is hidden; the Blondie writes the name of the grave where the gold is hidden on a stone that he puts in the center of the courtyard; then, he challenges Tuco and Angel Eyes to a duel (a *"truel"*) among the three of them, the one who will survive will take the gold. There is now a triangle, made up of the Good, the Bad, and the Ugly; each one of them must shoot the others, but each one can fire only one shot at a time; targeting one opponent, unavoidably he gives time to the second opponent to kill him. This triangle is the typical example of the so-called *Mexican standoff*, say a confrontation in which no strategy exists that allows any party to achieve victory. The reader who did not see the movie, or who did not remember it, could fruitfully search the sequence on the Internet[3], where the three men stare at each other during 10 minutes of silence, glances, and close-ups. It is one of the most epic showdowns in film history; a real masterpiece punctuated by Ennio Morricone's great music. Indeed, assuming that (1) each gunman—A, B, and C—has the same probability of hitting his target; (2) each gunman may shoot a limited number of bullets; and (3) the three gunmen might shoot simultaneously or sequentially, and who shoots first will kill his target; the first who shoots has 100% odds to die. In fact, if A shoots B or C, the survivor has time enough to kill him; and so on. The sole strategy to half the risk is to wait deliberately that someone else shoots, and then to kill him (provided that the target was the other opponent). If all the three players know the trick, they enter a loop. No one fires first, and the "truel" will never start, they will remain paralyzed looking at each other forever.

Jacques Lacan, the French psychoanalyst, discussed a situation very similar in his 1945 essay *"Logical Time and the Assertion of Anticipated Certainty: A New Sophism"* [5]. There are three prisoners—tells Lacan—a prison warden announces them that he will free the one who win a challenge. The prison warden has five disks differing only in color, three white and two black disks. He will fasten one of them between the shoulders of each prisoner without letting know which he selected. Prisoners cannot look the

---

disk on their shoulders; they can only see the disk between the shoulders of their fellows. By considering their companions and their disks, each prisoner must infer whether he is black or white. The first who will deduce his color must move toward the door, and he will be free. The situation is the reverse of the *Mexican standoff* because in this game all the three players unavoidably win (at least in principle). In fact, each prisoner could see either (1) two white disks; (2) one white and one black disk; or (3) two black disks. Knowing that there are three white and two black disks, if the first prisoner sees that his companions have two black disks fastened, he will immediately and unerringly deduce that he has a white disk and he goes to the door. If he sees that his companions have one black and one white disk, he could be either black or white. Nevertheless, if he were black, one of his companions would see two black disks; consequently, one of them would immediately go toward the door because he would know to be himself white. Therefore, if the first prisoner sees one of the two companions running toward the door, he can immediately do the same, because it means that he is black. Similarly, if both companions hesitate, it means that they both see two white disks, then he is white; he can thus go to the door. In conclusion, given that each prisoner can carry out the same reasoning, all three players might go simultaneously to the door. This game reveals the trick, which allows solving also the *Mexican standoff*. Both dilemmas can be solved only by using anticipation or retroaction of time; in other words, each player must guess the move of the other two opponents before deciding his move. This operation is precisely what machines cannot do. To be sure, machines can reckon the odds of other players according to the course of action they opt for, as well as they can calculate what decision would optimize the probabilities to win; finally, machines, provided with advanced sensors, can also detect early signs of any decision taken by others, by processing subliminal signals, which would go unnoticed to human eyes; consequently they might react before any human could do. Brief, there is no doubt that in real life, an intelligent machine could shoot, or run toward the door, before its fellow gunmen or prisoners, but it would need a material input which might show (even subliminally) the decision taken by other players. If human opponents do not reveal their intentions, not even through subliminal signals, or if the machine is opposed to other intelligent machines, programmed to win and with the same calculation power, the situation is a stalemate. Machines cannot create imaginary situations in which they think *as they were someone else* [6]. Intelligent machines can play quite well Bridge tournaments, but they are bad in bidding and the initial phases of this card game [7]. Indeed, during the bidding stage, each player knows only his/her cards and must play to guess at the hidden hands. This guess is psychologically possible by anticipating and retroacting the conclusions reached by other players. In other words, each player decides his bid by conjecturing past (how were the cards dealt among players?) and future (what will the play of other players be?) situations and projecting them on the critical instant of his decision. This instant can be conceptualized as a point on which all temporal lines, past and future, converge and intersect each other: it is the right moment to decide. Also, the gunman who decides whether to shoot or the prisoner who decides whether to get out needs to catch the right moment. We enter thus a different dimension of time, stretched between waiting and haste, hesitation and urgency. This time is no longer chronological time; it is a clot of tension, which can explode at any moment. The lesson that Lacan draws from the sophism of the three prisoners is thus that there is a temporality without reference to the clock; a mental temporality, which is not simply subjective, e.g., psychological time, but it is objective, as long as it is shared by all humans and it is articulated in a logical, nonspatial, structure. Within such a temporality, Lacan distinguishes three distinct moments: (1) the instant of the gaze, (2) the instant of the comprehension, and (3) the instant of the decision. Each one of these three moments is the expression of a logical, not chronological, punctuation. In other words, they are not a form of duration but are pure, immediate intuitions. The three gunmen, as well as the three prisoners, may reach simultaneously the same conclusions because these three moments are not chronological units, rather they are "*the intersubjective time that structures human action*" [8]. They are examples of the time opportune, say, the kairos.

To operate, e.g., to take a train, convene a meeting, cook a pizza, and so, we must think of the time in spatial terms, as it were made up by a chain of equal moments, from the past, through the present, into the future; otherwise, we will always miss trains and meetings, and burn pizzas. Nevertheless, human mental time is not made up of *equal moments*, rather of *meaningful instants*: "Now" which are unique, singular, "atoms of sense." Argentinian writer, essayist, and poet, J. L. Borges, superbly expressed this concept in his poem *Doomsday*:

"No hay un instante que no pueda ser el cráter del Infierno/ No hay un instante que no pueda ser el agua del Paraíso/ No hay un instante que no esté cargado como un arma/ En cada instante puedes ser Caín o Siddharta, la máscara o el rostro / En cada instante puede revelarte su amor Helena de Troya / En cada instante el gallo puede haber cantado tres veces / En cada

instante la clepsidra deja caer la última gota." (J. L. Borges, Doomsday, in *Los conjurados,* Buenos Aires, 1985)".[4]

The notion of "now" is extraneous to modern science; fundamental laws of physics ignore it, and there is no experimental way to establish it. According to an anecdote told by Rudolf Carnap, once Einstein said that *"the problem of the Now worried him seriously. He explained that the experience of the Now means something special for man, something essentially different from the past and the future, but that this important difference does not and cannot occur in physics. That this experience cannot be grasped by science seemed to him a matter of painful but inevitable resignation. I remarked that all that occurs objectively can be described in science; on the one hand, the temporal sequence of events is described in physics; and, on the other hand, the peculiarities of man's experiences with respect to time, including his different attitude towards past, present, and future, can be described and (in principle) explained in psychology. But Einstein thought that these scientific descriptions cannot possibly satisfy our human needs; that there is something essential about the Now which is outside the realm of science"* [9].

A further formulation of the three-prisoner and the *Mexican standoff* dilemmas is known in logic as the Buridan's ass paradox, named after the 14th century French philosopher Jean Buridan. The paradox reads: *"An ass placed equidistant between two bales of hay must starve to death because it has no reason to choose one bale over the other [...] The general principle underlying the starvation of Buridan's ass can be stated as follows:* a *discrete decision based upon an input having a continuous range of values cannot be made within a bounded length of time. Buridan's ass starves because it cannot make the discrete decision of which pile of hay to eat, a decision based upon an initial position having a continuous range of values, within the bounded length of time before it starves. A continuous mechanism must either forgo discreteness, permitting a continuous range of decisions, or must allow an unbounded length of time to make the decision"* [10]. Ironically enough, intelligent machines tend to behave such as the stupid ass of the tale. In electronics, the metastability problem, or arbiter problem

---

[4]Any instant can be the hell crater/Any instant can be the heaven water/Any instant is loaded as a gun/At every instant, you can be Cain or Siddhartha, mask or face/At every instant, Helen of Troy may reveal her love for you/At every instant, the rooster may have crowed three times/At every instant, the hourglass is about to drop the last drop.

(from the device, *arbiter*, used to face it[5]), is indeed a real-life application of the Buridan's paradox. Metastability is a condition in which the circuit pauses, becoming incapable of making any option. Metastability can arise when reading asynchronous inputs, generated by a computer interacting with an external device, an interaction requiring analogue to digital conversion, multiple clock domains on the same chip. In a metastable condition, a signal is sampled close to a transition, leading to indecision as to its correct value, *"for example, if a 0 is represented by a zero voltage and a 1 is represented by +5 volts, then some wire might have a level of 2.5 volts. This leads to errors because a 2.5-volt level could be interpreted as a 0 by some circuits and a 1 by others"* [10]. A well-designed arbiter can ensure that all delays are very brief, but it cannot eliminate the problem. In other words, although metastability can be practically dealt with, no technical fix can prevent it, because it is structurally connected with how digital machines are built to cope with time and decisions.

4.4 What Computers Can't Do

AI incapacity to perform some functions and activities that are standard for natural intelligence is the basis for the rejection of the very notion of intelligent machines. Radical critics of AI programs argue that the concept of AI is misleading. To be sure, computational machines exist, they are helpful, and they can improve human performances. However, when we state that a computer *"calculates,"* we are unknowingly using a metaphor, as when we state that a telescope and a microscope *"see."* Humans calculate *through* computers and see *through* telescopes and microscopes[6]. All technology devices do not autonomously act even when they are highly automated, but they improve more and more hugely human ability to do something; they are not agents, they are tools. This argument was first raised in the 1970s by Berkeley philosophy professor, Hubert Dreyfus. His book, "What computers can't do" [11], is paramount for all those who are interested in machine intelligence. Dreyfus argues that *"facts are not relevant or irrelevant in a fixed way, but only in term of human purposes, (...) Since a computer is not in a situation, however, it must treat all facts as possibly relevant at all times.*

---

[5] An arbiter has two stable states corresponding to the two choices, each request pushes the circuit toward one stable state or the other.

[6] This comparison does not consider, however, the difference between digital and analogue devices; this difference has important consequences that I will illustrate in the next chapters.

This leaves AI workers with a dilemma: they are faced either with storing and accessing an infinity of facts or with having to exclude some possibly relevant facts from the computer's range of calculations" (11, p. 258). His argument is that machines are programmed as though the world were made by atomic, out-of-context, facts, governed by formal rules[7], which is a model good enough to perform some tasks, but incapable of explaining the complexity of the human world. Dreyfus' point is the old philosophical issue[8] about how the world could be simultaneously seen as made up of both continuous and discrete quantities, the same problem which generates the Buridan's ass paradox. Dreyfus argues that machines can operate only on discrete, context-free facts; consequently, they are obliged to turn continuous quantities, which have an infinite number of steps, into finite and countable data.

On the contrary, human intelligence operates through continuous elements. Although we can isolate atomic facts—such as figures in the foreground—we always perceive the background and unconsciously we interpret the foreground through, and thanks to, the background. In other words, human beings always perceive the world as a gestalt, structured by their intentions and purposes: "*The human world, then, is prestructured in terms of human purposes and concerns (...) This cannot be matched by a computer, which can deal only with universally defined, i.e., context-free, objects. In trying to simulate this field of concern, the programmer can only assign to the already determined facts further determinate facts called values, which only complicates the retrieval problem for the machine*" [11, p.261]. To Dreyfus, this gap is not a *transitory* limitation, destined to be overcome by technology progress, but it is a structural limit inherent to any machinery, no matter how sophisticated it is. This limit makes misleading the notion of AI, because—he argues—machines will never be able to operate with concepts such as situation and purposes, and they will never develop the holistic vision necessary to perform activities such as learning a natural language or successfully competing with a human chess master. So, although Dreyfus' arguments were convincing, his conclusions were wrong.

Dreyfus grasped something important—as even some computer scientists later admitted [12, 13]—but he was misled by AI of first generation, the

---

[7]Dreyfus calls this assumption "psychological assumption," to distinguish it from biological, epistemological, and ontological assumptions, which concern respectively how the brain is organized, the structure of human knowledge, and the configuration of real world.

[8]The problem was first debated by the pre-Socratic Greek philosopher, Zeno of Elea (c. 495 – c. 430 BC), and the Eleatic School.

so-called "*Good Old-Fashioned AI*" (GOFAI). "GOFAI" was constructed on high level symbol manipulation, assuming that "*although human performance might not be* explainable *by supposing that people are actually following heuristic rules (. . .), intelligent behavior may still be* formalizable *in terms of such rules and thus reproduced by machine*" [11, p. 189]. GOFAI devices were just sophisticated computational machines. Despite their sophistication, they were only tools for automating calculations and operating with formal logical symbols. In 1992, when Dreyfus wrote the introduction to the MIT edition of his book [14], it was already apparent that the new generation of AI devices was instead game changing. The rapid increase in processing capacities and speed, coupled with exponential growth in data volumes and storage, algorithmic improvements (e.g., evolutionary algorithms, genetic algorithms, swarm intelligence algorithms, and so on), advances in machine learning and perception, new "statistical learning" techniques such as hidden Markov models and neural networks, created a new scenario. Next-generation AI was beginning to learn by seeing, reading, viewing, watching, and searching. Not only Dreyfus' criticisms were disproved by technology progress, but the very theoretical foundation of his criticism was shifting. Dreyfus, then, posed the ultimate objection to AI: "*all work in AI, then, seems to face a deep dilemma. If one tries to build a GOFAI system, one finds that one has to represent in a belief system all that a human being understands simply by being a skilled human being (. . .) Happily, recent research in machine learning does not require that one represent everything (. . .) But then, as we have just seen, one encounters the other horn of the dilemma. One needs a learning device that shares enough human concerns and human structure to learn to generalize the way human beings do. And as improbable as it was that one could build a device that could capture our humanity in a physical symbol system, it seems at least as unlikely that one could build a device sufficiently like us to act and learn in our world*" (Dreyfus 1992, xvi). Dreyfus' point is thus a phenomenological and existentialist objection to machine intelligence: machines cannot have actual experiences, because they do not have intentions, motivations, volitions; they cannot love or hate, they cannot feel happy or unhappy.

The main weakness of the phenomenological objection is that it takes as granted what it should instead demonstrate. There is no doubt that, for now, current technology devices do not have "experiences" in the human sense of the term, but could we exclude that they will be ever able to develop such an ability? Already today, machines can do most activities that, in 1972, Dreyfus thought to be impossible to them. Natural language processing approaches are developing the capacity of encoding semantic commonsense knowledge and machines promise to acquire narrative skills soon. A variety of next-generation sensors are dramatically improving machines' capacity for sensing the environment, including increasing capabilities for speech, facial, and object recognition. *Nouvelle artificial intelligence* creates robots provided with embodied minds, which learn through the inputs they receive from the external world. New generation algorithms, based on fuzzy and paraconsistent logics, allow creating subsymbolic and nongoal systems, which can "learn from the experience" and fix their goals autonomously, only according to the training data. Intelligent machines can increasingly recognize, interpret, process, and simulate human affects. Brief, the history of AI shows that one must be very cautious fixing theoretical limits to technological advances. As a Heideggerian philosopher, like Dreyfus, should have known, technology is only ruled by the will for a boundless increase of power.

4.5 Esprit de Géométrie and Esprit de Finesse

It seems thus that we have reached a dead end; as much as objections to the notion of AI are reasonable, they are—and seem to be destined to be—disproved by technology progress. Ultimately, we are still far from being able to answer the question of whether AI will be ever able to simulate, maybe surpass, human intelligence.

Computational machines, as we know them today, were invented in the 17th century in France by Blaise Pascal, the French scientist and philosopher. In 1642, the young Blaise was 19. He was obliged to spend his days, sometimes even nights, helping his father, a tax collector for Upper Normandy, in interminable, grueling, calculations of taxes. He was probably bored to tears, so he devoted himself to a way to get free from this tedious task. Being a genius, instead of telling off his dad, Pascal invented the first digital calculator of human history. Pascal's calculator—also known as the arithmetic machine or Pascaline—is still today the paradigm of all computational devices.

This story shows well what automation is for. Be Pascal's calculator, robots in assembly lines, currency exchange machines, or HAL 9000; it is always the same. The more a task is boring and unrewarding, e.g., calculating taxes, assembling pieces of an item, exchanging currencies, and calculating a spaceship route, the more we try to dump its burden on machines. Computational devices are for automating boring and repetitive activities and increasing the speed of their execution, including those mental activities which are less gratifying such as, e.g., solving a diophantine equation in 26 variables or identifying people by confronting their actual faces with photos on their passports[9]. These repetitive activities increased exponentially with the industrial revolution and, consequently, also computational devices spread and became ubiquitous. When we speak of AI, we must thus consider that we automated only a fraction of natural intelligence, the one involved in wearisome and uninspiring activities, setting aside other functions and abilities. Natural intelligence includes a variety of skills, which have not been taken into consideration—at least up until now—by automation.

It was precisely Pascal to propose the distinction between *esprit de géométrie* and *esprit de finesse*. The *esprit de géométrie* is the analytic mind, which always distinguishes and dissects reality into elementary components. The *esprit de finesse* is instead intellectual finesse, the perception of those things that can't be dichotomized and analyzed, such as music, arts, religion, human affects and emotions, and the horrible and the sublime; it is the feeling of the whole and the unspeakable. *Esprit de géométrie* and *esprit de finesse* must not be thought as two different realms, rather they are the two sides of the same coin, which is human intelligence: "*We must see the matter at once, at one glance, and not by a process of reasoning, at least to a certain degree. And thus it is rare that mathematicians are intuitive, and that men of intuition are mathematicians, because mathematicians wish to treat matters of intuition mathematically, and make themselves ridiculous, wishing to begin with definitions and then with axioms, which is not the way to proceed in this kind of reasoning*" [15, p. 2].

Starting with Pascal, humans began to create machines (GOFAI included) to automate and expedite tedious operations relevant to the *esprit de géométrie*. By contrast, they did not invent devices to automate activities pertaining to the *esprit de finesse*. In fact, these activities can hardly be defined "boring"; rather, they constitute the most pleasant and attractive part of human mental life. Moreover, they are much less relevant to industrial production, and consequently less economically significant, than repetitive actions based on calculations, logical operations, procedures, and algorithms. Something radically new happened in the 1990s with the birth of the World Wide Web. The web has been increasingly offering to each one the possibility

---

[9]Of course, one cannot exclude that there are human beings who are pleased to calculate diophantine equations or to check passports, but most persons are probably happier if an intelligent machine does these activities for them.

of being always connected; using and producing contents; remotely making money and sex; gaming, trading, and flirting online; enjoying music and videos; overcoming spatial and temporal barriers; mixing languages and linguistic codes; sharing memories and knowledge; and so. These activities are emotionally and economically rewarding, but they require machines more and more able to recognize, emulate, and interact with human imaginative insight and conjectural skills. The online digital world aims to become—so to speak—a Turing's nightmare: a place where humans and machines are indistinguishable. The electronic world is highly immersive; the late Marshall McLuhan noticed that electronic communication is much more reactive and emotionally intense than any previous form of communication [16]. Walter Ong spoke of "second orality" or "electronic orality" [17] to describe the digital society. The digital, interconnected society needs machines, which can hybridize with human beings, precisely because humans and machines must become fully interchangeable online, to increase effectiveness, operational capacity, and economic profitability. The expressions *"digital unconscious"* has been recently used by various scholars to describe how a collective human–machine mindset is emerging, beyond the awareness of single users[10]. In 2010, Lydia H. Liu, W. T. Tam Professor in the Humanities at the *Institute for Comparative Literature and Society*, Columbia University and director of the *Center for Translingual and Transcultural Studies* at Tsinghua University, Beijing, devoted a scholarly book [21] to a new understanding of human–machine interactions at the unconscious level, based on the idea of an increasing symbiosis of the computing machine (and the digital world) and the human unconscious. Canadian sociologist, and former Marshall McLuhan assistant, Derrick de Kerckhove, conceptualized the "digital unconscious" as a collective human–machine intelligence emerging from the whole information shared online, which would arise *"from hybridization between real and virtual, marked by reduced interiority, connected to the self, and an extended externality linked to the networked world"* [22]. In 2015, Mireille Hildebrandt provided an extensive description of "digital unconscious": *"We are in fact, surrounded by adaptive systems that display a new kind of mindless agency. Brain-inspired, neurosynaptic chips have been prototyped, that are typical for the way long-existing technologies such as artificial neural networks and miniaturization of ever more integrated circuits on silicon chips combine to simulate one of the most critical capacities of living organisms:*

---

[10]At the beginning, the expression was chiefly used to mean only the huge amount of personal information unwarily shared on the web [18, 19, 20].

unconscious, intuitive and on-the-spot pattern recognition. The environment is thus becoming ever more animated. At the same time, we are learning slowly but steadily to foresee that we are being foreseen, accepting that things know our moods, our purchasing habits, our mobility patterns, our political and sexual preferences and our sweet spots. We are on the verge of shifting from using technologies to interacting with them, negotiating their defaults, pre-empting their intent, while they do the same to us. While the environment gets animated, we are reinventing animism, ready to learn how to anticipate those that anticipate us – animals, fellow humans and now also smart, mindless machines" [23, ix]. This situation has created—for the first time in history—the will to automate the *esprit de finesse*. In other words, as the industrial civilization needed to automate the *esprit de géométrie,* the digital culture is obliged to embark on the adventure of attempting to automate the *esprit de finesse.*

So, at this juncture, we can provide a first, provisional, answer to the initial questions, whether AI could ever take over natural intelligence and why this question is in the limelight. AI is in the limelight because it is driven by the epochal transformation from the industrial to the digital society. Intelligent machines, provided with something which emulates (or at least simulates) the *esprit de finesse,* are the critical technological shift needed to achieve such a transition. When powerful historical and economic forces push technology in a direction, there is no way to change this evolution. The fashion for machine intelligence and the flourishing of cultural narratives and imagery about AI—scary tales on the rebellion of machines included—express the pervasiveness and the hegemonic power of the material forces driving technology innovation. Single AI programs can fail, as it happened in the past, but AI is here to stay, notwithstanding any philosophical and scientific criticism. However, although there is no doubt that machine intelligence can outperform human intelligence as far as the *esprit de géométrie* is concerned, it remains highly controversial whether AI programs, aiming to duplicate the *esprit de finesse*, could be ever accomplished. The question of whether intelligent machines are destined to become intelligent agents, or whether they will remain forever what Chalmers called "philosophical zombies" [24] is still unsolved. Nevertheless, now this question can be reformulated in different terms, asking whether AI will be ever able to automate the *esprit de finesse*.

4.6 The Symmetrical Logic

At the beginning of this chapter, I have illustrated some situations in which intelligent machines are unable to emulate humans, e.g., understanding the bizarre behavior of a ring of money counterfeiters, get rid of the Mexican standoff, recognizing the kairos in the three-prisoner dilemma, and solving the Buridan's ass paradox. Human beings could cope with these situations by some forms of intuition and conjectural capacity, say, employing the *esprit de finesse*. In principle, machines provided with an automated *esprit de finesse* should be able to face them quite easily. What are thus the capacities that machines must develop to achieve such a result? All these situations are dealt with humans by resorting to their capacity for perceiving the whole and the now.

Whole and now apparently belong to two different realms. The whole seems to concern space, while the now seems to concern time. However, on a closer look, one can realize that whole may refer to time as well, *"To view the world sub specie* aeterni *is to view it as a whole - a limited whole. Feeling the world as a limited whole - it is this that is mystical"* [25, 6.45]. Similarly, the notion of now hides a spatial dimension. Scholastic philosophers distinguished between "the now that passes" (*Nunc fluens*), which is the ephemeral human time; and "the now that remains" (*Nunc stans*), which is the eternal time of God.[11] However, this distinction might create an unbridgeable gap between humanity and God, because the two now can hardly meet. In such a way, human beings and God would be destined to live in two separate, parallel, noncommunicating temporal dimensions, which are unbearable for religion, such as Christian religion, centered on the incarnation and the sacred history. So, Nicholas of Cusa, the great Renaissance humanist and philosopher, introduced a third distinction, the *Nunc instantis*, the "instantaneous now." The "instantaneous now"— argued Cusa—is the moment of eternity, which cuts the continuum of the chronological time. Each instant is potentially eternal because it can be indefinitely expanded in the subjective experience, yet it is not the eternity, because it has no duration at all, i.e., you cannot catch the instant. It is a point with zero dimensions but with a location determinable by an ordered set of spatiotemporal coordinates. Brief, the *Nunc instantis* is the eternity as human creatures can experience it, and it is the place where they can get in

---

[11]This distinction was first introduced by Boethius in The Consolation of Philosophy.

contact with God[12]. Similarly, the last century German philosopher, Walter Benjamin, argued that human time is not homogenous, but it is *"fulfilled by the here-and-now (Jetztzeit)"* [26]. Instantaneous now, here-and-now, are—according to Italian philosopher Massimo Cacciari—the door which unites two sides by separating them: the present that urges and the indifferent eternity [27, p. 52]. The whole and now are thus almost the two sides of the same coin. They are how humans speak of two unthinkable concepts: infinity and eternity.

Thinking the unthinkable is a unique feature of the human mind; this is the fundamental discovery of influential Chilean psychiatrist and psychoanalyst, Ignacio Matte Blanco [28]: *"Deep down, both the infinite and the unconscious are human attempts, independent of one another, at understanding something which is indivisible and, as such, unthinkable"* (Matte Blanco 1975, p.377). Matte Blanco was one of the most eminent psychoanalytic scholars of the second half of the 20th century. With his ground-breaking research, he attempted to formalize the theory of the unconscious using the formal logic of Russell and Whitehead. *"The most important general conclusion that emerges from my studies—Matte Blanco writes—is that psychical life can be viewed as a perceptual dynamic interaction—in terms of tension, cooperation, or even union—between two fundamental types of being which exist within the unity of every man"* [30, p. 13]. Matte Blanco argues that the human mind has two fundamental operational modes[13], with their respective logics: (1) the classical logic, in which each element is defined by spatiotemporal coordinates that he calls asymmetrical and (2) the logic of totality, atemporal and spaceless that he calls symmetrical.

---

[12]Nicholas of Cusa is thus arguing that we can get in contact with the transcendent through the instant. This idea is nicely expressed also in Sufi stories and Zen koans, such as the Zen parable of the man, the tiger, and the strawberry: "A man traveling across a field encountered a tiger. He fled, the tiger after him. Coming to a precipice, he caught hold of the root of a wild vine and swung himself down over the edge. The tiger sniffed at him from above. Trembling, the man looked down to where, far below, mother tiger was waiting to eat him. Only the vine sustained him. Two mice one white and one black, little by little started to gnaw away the vine. The man saw a luscious strawberry near him. Grasping the vine with one hand, he plucked the strawberry with the other. How sweet it tasted!" [29, p. 38].

[13]Similarly, Nobel laureate Daniel Kahneman suggests that there are two modes of thought; "System 1," fast, instinctive, and emotional; "System 2" slower, deliberative, and logical [31].

Asymmetrical logic is the logic followed by conscious thought, which allows the conception and perception of concrete and well-delimited things, such as a person, an object, a thought referring to a concrete fact, a single abstract concept, and so on. Asymmetrical logic is governed by the principle of noncontradiction, e.g., A cannot be the same as non-A, $A \neq non\text{-}A$. In Pascal's terms, asymmetrical logic corresponds to the *esprit de géométrie*.

Symmetrical logic is instead the sense for what cannot be broken down, the feeling for the whole [32]. Symmetry is a state of no limits, where the unit person feels as one with the world outside and inside, it is the deepest part of our mind, its primordial unconscious structure, which "*does not know individuals but only classes or propositional functions which define the class*" [30]. Symmetrical logic is characterized by (1) absence of contradiction so that any assertion is equal to its negation; (2) absence of distinction between mental and external reality; and (3) absence of boundaries in time and space (eternity and infinity). Symmetrical logic is made up of sets of infinite sets and is governed by the principle of totality, e.g., A is the same as non-A, although at a further dimensional level, $A(x) \Leftrightarrow non\text{-}A(x)$. Ultimately, with the term "symmetrical logic," Matte Blanco provides his description of the *esprit de finesse*.

The coexistence of symmetrical and asymmetrical logics, which is called bi-logic by Matte Blanco, is the standard condition of all human beings: "*we live the world as though it were a unique indivisible unit, with no distinction between persons and / or things. On the other hand, we usually think of it in terms of bi-logic and, some few times, in terms of classical logic*" [30, p. 46].

Now it is possible to reformulate more precisely the question of whether we can ever accomplish to automate the *esprit de finesse*. The point is whether we can ever design intelligent machines capable of applying symmetrical logic.

4.7 Data and Totality

In the previous chapters, I have extensively argued that capacities for intuition largely coincide with a particular feature of human intelligence, say, the capacity for operating through indeterminate concepts such as infinity, eternity, totality, instantaneity, and so on. I called such a capacity the capacity for "*thinking the unthinkable*," say, the *indivisible,* the whole. Following Matte Blanco, I argued that this capacity can be comprised under the wider category of symmetrical logic. With Matte Blanco, I argued that symmetrical logic is not a marginal aspect of the human mind, rather it is the basic and primordial mental skill that makes us humans. Most creations of the human spirit—such as arts, music, poetry, mystics as well as sense of humor, comic timing, perception of the kairos, the coup d'oeil, and so on—are the result of symmetrical logic.

It is thus understandable why computer scientists and AI researchers have been investigating for several years nonclassical logic systems with the aim to create machines capable for "*thinking like human beings.*" Logic systems can be categorized under two main headings: (1) explosive and (2) nonexplosive. Explosive systems are systems based on the principle *ex falso (sequitur) quodlibet* (EFQ), "from falsehood, anything (follows)"; they affirm that from a contradictory statement one can infer any conclusion; i.e., if one accepts paradoxes and antinomies, it is impossible to reach any truth. "*Classical logic, and most standard 'non-classical' logics too such as intuitionist logic, are explosive. Inconsistency, according to received wisdom, cannot be coherently reasoned about*" [33]. On the contrary, nonexplosive (paraconsistent) systems admit contradictions; paraconsistent logic does not deny the notion of truth, rather it claims that one can reach true conclusions even from contradictory premises. It is out of the scope of this chapter to describe the various paraconsistent systems and the techniques that they use, it is, however, important to stress that they do not truly accept antinomies and paradoxes (as one could erroneously suppose) rather they try to neutralize contradictions by including—instead of excluding—them [34, 35]. "*Paraconsistent logic accommodates inconsistency in a controlled way that treats inconsistent information as potentially informative*" [33].

Next-generation AI is increasingly applying algorithms based on nonclassical logic systems; e.g., fuzzy set theory, computability logic, interactive computation, and so on [36, 37]. This approach is producing some results and machines are starting to simulate human capacity for intuitive thinking. On the basis of the past experience, one can foresee that AI will become more and more skilled in imitating human intuition, and it is likely that next generation AI will become eventually able to replicate human intuitive capacities almost perfectly. Yet, algorithms using paraconsistent logic are not based on actual symmetrical processes. Paraconsistent algorithms deal with inconsistent information by turning it into binary, dichotomic information. At the end of the day, paraconsistent algorithms are a tool to incorporate antinomies within classical logic systems, allowing machines to handle contradictory concepts [38]. When symmetrical notions—such as infinity, eternity, totality, and instantaneity—are handled by AI, they are eventually included into binary, dichotomic processes.

Next-generation AI can put up with *A(x)* ⇔ *non-A(x)* notions, only provided that they are incorporated into *A* ≠ *non-A* terms. One could argue that this is due to the still limited development of AI, it would be thus conceivable a future in which new, more advanced algorithms will be able to operate through actual symmetrical operations, say, only with indivisible and indeterminate notions. I disagree with this hypothesis, as computational machines capable of doing completely without computing are an oxymoron, a *contradictio in adiecto*. One could still argue that there is a threshold above which quantity may turn into quality; a myriad of details may become the whole picture; and the *esprit de géométrie* may become *esprit de finesse*. I answer that this is not possible because of data. What is data?

Data is a difference. According to Floridi "*a datum is reducible to just a lack of uniformity* (diaphora *is the Greek word for "difference"), so a general definition of a datum is* The Diaphoric Definition of Data (DDD)*: A datum is a putative fact regarding some difference or lack of uniformity within some context. Depending on philosophical inclinations, DDD can be applied at three levels: (1) data as* diaphora de re, *that is, as lacks of uniformity in the real world out there (...) As "fractures in the fabric of being" they can only be posited as an external anchor of our information (...); (2) data as* diaphora de signo, *that is, lacks of uniformity between (the perception of) at least two physical states, such as a higher or lower charge in a battery, a variable electrical signal in a telephone conversation, or the dot and the line in the Morse alphabet; and (3) data as* diaphora de dicto, *that is, lacks of uniformity between two symbols, for example the letters A and B in the Latin alphabet*" [39]. The DDD thus implies that one can never represent indeterminate concepts (e.g., infinity, eternity, totality, and instantaneity) by using data. On the one hand, data is an asymmetry, a fracture, a lack of uniformity, a difference; on the other hand, infinity, eternity, totality, and instantaneity—the whole and the now—cannot admit dichotomic divisions and internal fractures. One can, of course, operate though discrete operators which symbolize indeterminate concepts (as it happens in mathematics), but in so doing one turns them into determinate quantities. Taken rigorously, indeterminate concepts cannot be handled by using data, because the very notion of data denies the existence of indivisible, indeterminate realities. If an item can be expressed in terms of data, it cannot be simultaneously, from the same account, expressed also in terms of totality. The indefinite cannot be generated by a finite collection of particularities as well as an infinite set—even a countable infinite set—is not made up of finite elements: by summing up all singularities, you will never generate the whole. Eventually, infinite sets cannot be reduced entirely to data; there is an unbridgeable gap between these two dimensions[14]. This is also the reason we cannot solve the metastability problem by the roots. Intelligent machines—even quantum computers—are devices which operate through data, more and more data as they become more and more intelligent. To think of a computational machine provided with symmetrical logic (i.e., a machine which can do without data) is therefore an inherent contradiction, no matter how sophisticated the machine is[15].

4.8 The Death of the Pythia

In 1976, Friedrich Dürrenmatt wrote the short novel *Das Sterben der Pythia* (The Death of the Pythia). The story unfolds as a dialogue between two characters, Pannychis XI, an elderly Delphic priestess at the end of her life, and Tiresias, the clairvoyant. Both cynic and unbelievers, nevertheless they are genuinely different, Pannychis "*wanted to use her oracles to mock those who believed them*"; Tiresias had a hidden political agenda instead to achieve. Their conversation was about the story of Oedipus. Although in different moments, they were both asked three times to unravel to Oedipus his fate. Three times they invented. Pannychis "*with imagination, with whimsicality, with high spirits, even with a virtually irreverent insolence, in short: with blasphemous jocularity*"; Tiresias "*with cool reflection . . . , with incorruptible logic, again in short: with reason.*" Ironically enough, all of Pannychis's implausible oracles turned out to become a reality; Tiresias' manipulative and intelligent predictions had the opposite of the intended effect. In the end, Tiresias says to Pannychis: "*Both of us faced the same monstrous reality, which is as opaque as man, who creates it.*" Pannychis does not answer; she fades away.

Dürrenmatt's short novel is almost a parable of AI research. No computational machine can truly duplicate the human mind in its entirety. Computer scientists and AI researchers who are using their "*incorruptible logic*" to develop AI capable for human intuitive skills are going only to imitate these capacities, creating soulless replicas rather than novel Adams.

---

[14] My conclusions are not that far from those of Dreyfuss, but my argument is quite different. I don't argue that machines will be never able to experience the world as a gestalt, I only contend that a computational machine provided with symmetrical logic is a nonsense.

[15] I don't exclude that in the future we could create biological computers using biomolecules; these artificial devices might be able eventually to reproduce symmetrical logic processes, yet they would not be any longer "machines," say mechanical devices, rather they would be "artificial biological organisms."

Machines replicating human intuition are not *super humans* but, so to say, counterfeit humans. Practically speaking, however, I will admit that they might work well enough to meet the main needs of the digital society, which basically consist in providing increasing interchangeability between humans and machines. Already today, bots can interact online with customers in an acceptable way, sometimes also challenging the Turing test. Eventually, even if machines will not become actual intelligent agents, their condition of "philosophical zombies" will be still workable for doing business. What is certain is that the debate on whether AI will ever surpass and substitute human intelligence is without merit. AI can do many things better than humans, but to use symmetrical logic, developing an *esprit de finesse.*

We are in the midst of an epochal transition only comparable to the transition from orality to literacy [40, 17]. AI is heralding this revolution, making it possible. I am convinced that AI will transform the labor market and overturn many current standards [41], and it will become more and more capable for miming some human intuitive skills. Yet, AI is great as far as it is used as a tool to amplify and enhance human analytic, dichotomic skills; this is its core mission.

AI is not for understanding the love for stage of a bunch of Sorrento counterfeiters, the paintings of the *Bamboccianti*, Sergio Leone's movies, and Morricone's music (not even to detect emotions in humans or to perceive the kairos in a critical decision). These things belong to a different register which is destined to remain forever extraneous to intelligent machines. Human reality is much more complex and richer than any computational device can grasp. Rationality is wider than we are used to think, there are forms of rationality—as Pascal teaches—that are understood by using intellectual finesse rather than computational capacities. The "*unthinkable*" and the "*unspeakable*" are not irrational, rather they express different—maybe higher—forms of rationality. It is not within the scope of this chapter to discuss whether this is good or bad; it is a fact and that's enough. "*There are, indeed, things that cannot be put into words. They make themselves manifest*" [25, 6.522]. Pannychis XI faded away, "*the rest is silence.*"

References

[1] ABI Research (2019). "Artificial Intelligence Meets Business Intelligence (White Paper)." ABI Research. 27 Sept. https://go.abiresearch.com/lp-artificial-intelligence-meets-business-intelligence.

[2] Kurzweil, R. (2005). "The Singularity Is Near: When Humans Transcend Biology". New York, NY: Viking Press.

[3] Bostrom, N. (2014). "Superintelligence: Paths, Dangers, Strategies". Oxford, UK: Oxford University Press.

[4] Huchel, B. (2019). "Humor is both a hurdle and a gauge to improve AI, human interaction." Phys. Org. 26 Feb. https://phys.org/news/2019-02-humor-hurdle-gauge-ai-human.html.

[5] Lacan, J. (1966). "Le temps logique et l'assertion de certitude anticipée. Un nouveau sophisme." In Écrits, by Jacques Lacan, 209. Paris: Seuil.

[6] Ventos, V., Costel, Y., Teytaud, O. and Thepaut Ventos, S. (2017). "Boosting a Bridge Artificial Intelligence." 2017 IEEE 29th International Conference on Tools with Artificial Intelligence (ICTAI). 1280–1287.

[7] Bethe, P.M. (2009). "The State of Automated Bridge Play." https://www.semanticscholar.org/paper/The-State-of-Automated-Bridge-Play-Bethe/8b229c32a3bcc983e5cf8d8cfbf743ec9a756a0b.

[8] Lacan, J. (1977). "Écrits: A Selection". Translated by Sheridan. Alan. London: Tavistock Publications.

[9] Carnap, R. (1999). "Intellectual autobiography". Volume 11 (Library of Living Philosophers). In: "The Philosophy of Rudolf Carnap", R. Carnap and P.A. Schilpp (eds.), pp. 3—84. Chicago: Open Court.

[10] Lamport, L. (2012). "Buridan's principle." Foundations of Physics, 42(8), 1056–1066.

[11] Dreyfus, H.L. (1972). "What Computers Can't Do. The Limits of Artificial Intelligence". New York, NY: Harper & Row.

[12] Crevier, D. (1993). "AI: The Tumultuous Search for Artificial Intelligence". New York, NY: Basic Books.

[13] McCorduck, P. (2004). "Machines Who Think". Natick, MA: A. K. Peters, Ltd.

[14] Dreyfus, H.L. (1992). "What Computers Can't Still Do. A Critique of Artificial Reason". Cambridge, MA: The MIT Press.

[15] Pascal, B. and Eliot, T.S. (1958). "Pascal's Pensees: The Misery of Man without God". New York: E. P. Dutton & Co., Inc. [accessed Nov 20, 2017]. https://www.gutenberg.org/files/18269/18269-h/18269-h.htm.

[16] McLuhan, M. (1970). "From Cliché to Archetype". New York: Viking Press.

[17] Ong, W.J. (1982). "Orality and Literacy. The Technologizing of the Word". New York: Routledge.

[18] deKerckhove, D. (1995). "The Skin of Culture: Investigating the New Electronic Reality". Toronto, ON: Somerville House Books.

[19] Monk, J. (1998). "The digital unconscious." In: "The Virtual Embodied: Practice, Presence, Technology", pp. 30–44. London: Routledge.

[20] Poster, M. (2006). "Information Please: Culture and Politics in the Age of Digital Machines". Durham, NC: Duke University Press Books.

[21] Liu, L.H. (2010). "The Freudian Robot. Digital Media and the Future of the Unconscious". Chicago: University of Chicago Press.

[22] Meckien, R. (2013). "The centrality of technology in the contemporary world." Institute of Advanced Studies of the University of São Paulo. [accessed 25 October 2018]. http://www.iea.usp.br/en/news/the-central ity-of-technology-in-the-contemporary-world.

[23] Hildebrandt, M. (2015). "Smart Technologies and the End(s) of Law. Novel Entanglements of Law and Technology". Cheltenham, UK: Edward Elgar Publishing Limited.

[24] Chalmers, D. (1996). "The Conscious Mind: In Search of a Fundamental Theory". Oxford, UK: Oxford University Press.

[25] Wittgenstein, L. (1922). "Tractatus Logico-Philosophicus". Translated by C.K. Ogden. London: Kegan Paul, Trench, Trubner & Co., Ltd.

[26] Benjamin, W. (1974). "On the Concept of History". Vol. I:2, in Gesammelten Schriften, translated by Dennis Redmond. Frankfurt am Main: Suhrkamp Verlag.

[27] Cacciari, M. (1994). "The Necessary Angel". New York: SUNY.

[28] Rayner, E. (1995). "Unconscious Logic: An Introduction to Matte Blanco's Bi-Logic and its Uses". London and New York: Routledge.

[29] Senzaki (2010). "101 Zen Stories". Whitefish: Kessinger Publishing LLC.

[30] Matte Blanco, I. (1975). "The Unconscious as Infinite Sets". London: Duckworth.

[31] Kahneman, D. (2011). "Thinking, Fast and Slow". New York, NY: Macmillan.

[32] Matte Blanco, I. (1998). "Thinking, Feeling and Being. Clinical Reflections on the Fundamental Antinomy of Human Beings and World". Howe, UK: Routledge.

[33] Priest, G., Tanaka, K. and Weber, Z. (2018). "Paraconsistent Logic", The Stanford Encyclopedia of Philosophy (Summer 2018 Edition), E.N. Zalta (ed.). <https://plato.stanford.edu/archives/sum2018/entries/logic-paraconsistent/>.

[34] Hewitt, C. (2008). "Large-scale organizational computing requires unstratified reflection and strong paraconsistency." In: "Coordination, Organizations, Institutions, and Norms in Agent", J. Sichman, P. Noriega, J. Padget and S. Ossowski (eds.), pp. 47–80. Berlin: Springer-Verlag.

[35] Buehrer, D.J. (2018). "A Mathematical Framework for Superintelligent Machines." ArXiv abs/1804.03301 n.pag.

[36] Cantone, D., Alfredo, F. and Eugenio, G.O. (1989). "Computable Set Theory." In: International Series of Monographs on Computer Science, Oxford Science Publications, xii, 347. Oxford, UK: Clarendon Press.

[37] Japaridze, G. (2009). "In the beginning was game semantics." In: Games: Unifying Logic, Language and Philosophy, O. Majer, A.-V. Pietarinen and T. Tulenheimo (eds.), pp. 249–350. Berlin: Springer.

[38] Abea, J.M., Nakamatsub K. and da Silva Filhoc J.I. (2019). "Three decades of paraconsistent annotated logics: a review paper on some applications". Procedia Computer Science, 159, 1175–1181.

[39] Floridi, L. (2019). "Semantic Conceptions of Information." The Stanford Encyclopedia of Philosophy (Winter 2019 Edition). 21 June. https://plato.stanford.edu/archives/win2019/entries/information-semantic/. Telecommunication Union (ITU), report on Climate Change, Oct. 2008.

[40] deKerckhove, D. and Viseu, A. (2004). "From memory societies to knowledge societies: The cognitive dimensions of digitization." World Report on "Building Knowledge Societies", UNESCO, Paris.

[41] Pew Research Center (2018). "Artificial Intelligence and the Future of Humans". Washington, DC: Pew Research Center.

5

Inclusively Designed Artificial Intelligence

Abhishek Gupta[1] and Jutta Treviranus[2]

[1]Montreal AI Ethics Institute, and Microsoft, Canada
[2]Inclusive Design Research Centre, OCAD University, Toronto, Canada

Abstract

Artificial intelligence (AI) can either automate and amplify existing biases, or provide new opportunities for previously marginalized individuals and groups. Small minorities and outliers are frequently excluded or misrepresented in population data sets. Even if their data is included, data-driven decisions favor the statistical average, thereby disadvantaging small minorities. Small minorities and people at the margins are also most vulnerable to data abuse and misuse. Current privacy protections are ineffective if you are an outlier or in some way anomalous. This chapter will discuss the challenges, dangers, and opportunities of machine learning and AI for individuals and groups that are not represented by the majority.

Keywords: Artificial intelligence, inclusive design, margins, disability, privacy, bias, diversity, transparency

5.1 Introduction

The data industry is currently one of the fastest growing industries globally, influencing every other industry and aspect of our lives. Among the fevered attempts to express the pervasive impact, data has even been called the new gold [1]. *The Economist* claims that data has surpassed oil as the most valuable resource [2]. There is a global race to dominate the field of artificial intelligence (AI) with numerous nations committed to achieving global leadership in AI, smart connected technologies, big data analytics, and other data sciences. Evidence in the form of data has become the yardstick for determining what is important and what is true in both the public and private sectors. Big data analytics, smart systems, and AI are disruptive technologies that are transforming how we make decisions, the very nature of work, and what is valued.

This fast moving power struggle comes with many challenges and opportunities. It presents an opportunity to either disrupt or amplify existing inequities. As more and more decisions are handed over to opaque automation, new and existing inequities and biases can become entrenched. The design of data analytics and machine learning carries a number of sources of bias. Among these include (a) human bias finds its way into the rules and algorithms in part due to the lack of representation among the engineers of the systems and in part because the data reflects the current biases in society, leading to biased algorithms derived from the data amplified and automated in the AI; (b) the data used to make decisions is skewed, with data gaps and under-representation of many minorities and marginalized groups, meaning they are not considered or recognized in the decisions; (c) the labels, tags, buckets, or classifiers of the data are not representative or discriminate and fail to fairly reflect marginalized groups; and (d) the most intractable problem is that by its very nature data-guided decisions will side with the majority or average in population-based decisions.

Similar to quantified statistical analysis before it, machine learning and AI are powered by statistical probability. This has a detrimental effect on small minorities and outliers, as probability is associated with the average or mean. Predictions and thereby decisions will be biased toward the majority or the average. A large homogeneous number is needed to sway the system in a given direction. Diversity, variability, and complexity do not meet these criteria.

People experiencing disability are an extreme example of this bias. Persons with disabilities are disadvantaged by these trends in that the data tools have many barriers to access; and, as a highly heterogeneous group, persons with disabilities are generally not recognized, understood, or served by data-driven systems. From a data standpoint, persons with disabilities are a collection of very small minorities that will not satisfy data processes that favor large homogeneous data sets, especially when making population data-based decisions. While people with disabilities are an extreme example, many other small minorities or digitally excluded groups experience similar bias.

5.2 The Illustrative Example of Disability

As an illustrative example of both the challenges and opportunities of data analytics, machine learning and AI, we will describe the scenario of people with disabilities. It can be said that persons experiencing disabilities are not only the most disadvantaged by the current design of AI, but also the group with the most compelling applications. People experiencing disabilities are very different from each other. The only common characteristic of disability is sufficient difference from the average that the design of most things does not work for you. In many data sets, individuals with disabilities will each be a minority of one.

Notions of representation do not hold when reflecting the needs of people experiencing disabilities. There is no clear, bounded set of characteristics when it comes to disability. Diagnostic categorization offers very few functional benefits, and risks further excluding people who don't fit or straddle the categories. For example, knowing someone is blind does not tell you whether the person is Braille literate, how much residual vision he/she has or uses, whether he/she once had sight and formed spatial models from the time he/she had sight, or any of the many other relevant characteristics of the person. As a result, most assumptions that come with disability categorization are flawed. To fully capture the relevant needs of all individuals that experience disabilities would require an unrealistically large number of representative groupings. For many people with disabilities, no one else can adequately represent their unique needs. This is an issue with data analytics and AI decisions in general.

At the same time, people with disabilities have the most to gain from technical innovation. They are frequently the technical pioneers that experiment at the bleeding edge, before the early adopters of new disruptive systems [3]. The majority of the technical innovations we take for granted were initiated in an attempt to address the challenges of disability, from e-mail, to telecommunications, to optical character recognition, text to speech and speech to text, to smart home systems. There is a saying that for most people technology makes things more convenient, for people with disabilities, it makes things possible. This also means that possibility is dependent on technology. If the technology ceases to work, the person may no longer be able to speak, write, move, or breathe.

Disabled people are also some of the most vulnerable to the abuses and misuses of data. Current privacy protections that resort to deidentification at source fail to protect disabled people. If you are highly unique, you can be reidentified. If you are the only person in a neighborhood to order a colostomy bag, for example, you can be reidentified. If you are the only person with a specialized wheelchair in your building, you can be reidentified. Most people with disabilities must barter their privacy for essential services. The act of requesting an exception, or something not usually provided, identifies you whether or not you provide your name, address, your government identification number, or whether or not your face is recognized. Data privacy for people that are outliers is often an unrealistic luxury that can't be protected by deidentification at source. At the same time, people with disabilities are often treated paternalistically or infantilized. They are not afforded the "dignity of risk." Systems are lacking that offer self-determination of the acceptable level of risk as a matter of informed personal choice and agency.

As a result of these challenges and opportunities, people with disabilities often inadvertently act as stress testers, and provide the early warning signs of the flaws of new systems. The flaws and risks in automated decision-making based on machine learning are significant. So are the opportunities if we use this disruption to transcend old patterns.

5.3 Bias in Data Collection and Reuse

One of the key areas where discrimination occurs starts right at the beginning of the machine learning development life cycle, namely in data collection [4]. This can be roughly split along two sources: (1) primary data collection and (2) subsequent reuse or secondary data use. Primary data collection is plagued by many different issues, the biggest of which stems from the use of convenience sampling [5] and assumptions around representativeness of the collected data to reflect the global reality in the context of the use of the machine learning system. Data is also often divorced from the context of collection and therefore is assumed to be context independent or transferable across contexts. The relevance and appropriateness of predictions or recommendations can be highly dependent on the context.

While a complete dive into the possible problems with this lie outside the scope of a brief chapter discussing inclusively designed AI systems, let us take a look at one facet of data collection, namely options presented during data collection that aim to aggregate people and their data into discrete fields. These disproportionately harm the outliers and individuals who have specific characteristics; e.g., individuals with minority needs that don't conform to the options presented. Often one of the approaches undertaken is to take a best guess at shoehorning data regarding these individuals into ill-suited categories. This practice creates inappropriate proxies for their actual characteristics. Another approach is to altogether ignore such individuals because they constitute "dirty data" that creates challenges in the downstream development of the machine learning system. What we essentially mean by this is that often in primary data collection we start with a set of fields that we're interested in but often those with minority needs have values for these fields that are missing, ill-defined, or require a new field which makes this stage challenging. Additionally, it places undue burden on the individuals who are marginalized to comply with requirements of this stage when they might be the ones that are most in need of the service.

A stark example of this is the new requirement to have an Aadhaar card (national ID in India) to obtain necessary medications for HIV [6]. The process to obtain this national ID [7] requires certain pieces of existing IDs which people from the most marginalized sections are often unable to produce. Thus, they face an undue burden [8] in being able to access services that are specifically geared to ameliorate their situation. In addition, even when they do have this ID, infrastructure failures [9] like biometric scanners that confirm the identity of the Aadhar card holder can fail. Because this is linked to a central digital system, the issuer of rations has no alternative but to deny rations to people that are most in need of them. Prior to the centralization and automation of the system, the ration issuer could use human judgement, guided by the intimate knowledge of the small communities served.

Data originally collected for one purpose is frequently used for further or secondary purposes. When public datasets are reused to guide other decisions there is often a lack of access to the original purpose of data collection, the assumptions and design choices of the classifiers or fields, how the data was collected, what pitfalls they faced, what data points were cleaned out and ignored, and many other critical factors that can compromise the resulting decisions. Additionally, the representation provided of the data might also skew downstream analysis ultimately infusing biases that can lead to unintentional and subtle harms. The labelling and fixed fields in that dataset and lack of access to the raw data will again disproportionately harm those who lie on the margins of what is considered to be normal data.

The issue of under-representation or misrepresentation in data is systemic and self-reinforcing. A particularly relevant example of the lack of representative data comes from the indigenous communities [10] that have a lower than average presence on the Internet and the use of digital technology. This means that their digital exhaust is very small and is not representative when digital exhaust is used as a common source of training data. Internet samples don't do justice to representing these communities and hence create a negative interaction experience, which further discourages the use of these systems by indigenous communities. This in turn makes for a vicious cycle where their use declines further because of this negative experience and further reduces the digital footprint. The same argument can also be made for other sections of society that have low Internet use due to digital disparity.

5.4 Bias in Data Processing

The data processing stage where decisions are made regarding how to transform, process, and prepare data to feed into a model, is another potentially important step that comes not only with both challenges but also corresponding opportunities to address the inclusive design of AI systems. Data brokers wishing to market data that can quickly and efficiently reach useful inferences for the majority, often emphasize dominant patterns at the expense of the minority. In addition to creating a bias against minorities and outliers, this results in a system that is less capable of addressing weak signals, the unexpected or a changing context.

One of the most common uses of a machine learning system is to use past data to produce predictions that are useful in contexts such as recommender services that suggest products on Amazon [11], movies on Netflix [12], songs on Spotify [13], for example. Here the primary harm is a winnowing of exposure to diverse perspectives and a popularity echo chamber where perspectives that have attention get more attention and perspectives that lack attention virtually disappear.

In the case of more mission-critical applications where decisions regarding significant aspects of human lives are made, the choices in how data is processed become very important. For those that fall into the average or mainstream as defined by the data collected, the processing and transformations are fairly benign as they retain the majority of the fidelity of the original and raw data and have a minimal gap between reality and the data which ultimately leads to more or less accurate predictions for them. But in the case of people that are represented as outliers, their interactions with the system can be highly skewed because they might not be sufficiently represented in the system, either because they were excluded at the data collection stage or they are so distant from the so-called average, as defined by the collected data, that the predictions made for them are mostly nonsensical or wrong. The normalization process used to optimize the data includes cleaning columns, filling default values where there are missing values, and other data manipulations. These processes result in data that is largely irrelevant or misrepresentative for outliers, thus harming how they are interpreted by the system.

These same data issues discussed above are passed on and complicated when performance benchmarks and benchmarked data are deployed in AI development. Even when the intention is to address a specific community challenge, when pretrained or off-the-shelf datasets that have been trained, validated, and tested on benchmark datasets are used, they miss out on the specifics of the communities that they are looking to serve with their machine learning system. Take for example natural language processing systems; some systems utilize Reddit's 1.65 billion comments dataset [14] to pretrain the layers of their model. While it constitutes a large corpus of text, one must analyze the source of the dataset and the underlying lack of diversity that is present. Reddit users form only a small part of the larger Internet-using community, even so, only a minority of users are responsible for posting a majority of the content, following roughly a power law as is the case for many online communities. To go from there, not all people and communities are equally represented in terms of Internet use and hence not on Reddit. Additionally, it is largely an English-speaking forum, thus further limiting the applicability of the system on a larger scale to users who don't utilize English in their interaction with the system. Many other benchmark datasets for language processing suffer from the same curse. We can also examine a similar issue in object recognition systems that utilize, e.g., the ImageNet dataset [15]. When it comes to the human category, there is a surprising lack of diversity in the kinds of people who form sample points in the dataset. Google Photos [16] notoriously suffered from confusing black people and gorillas in their autotagged images which stemmed from such lack of diversity in the underlying datasets.

These practices are very pervasive and there is minimal interest in improving the situation because often the affected individuals constitute a minority of the intended audience and the error rates that disproportionately affect them represent a very small part of the overall system accuracy, and thus, are tolerated in the service of rolling out production systems to optimize profit. As competitive corporations are leading the adoption of AI and automated decision, competitive business, rather than public values, is reflected in the design.

5.5 Bias in Buckets and Labels

The primary modus operandi of statistical and machine learning systems is to bucket people or objects of interest to make predictions about them. The smaller or more specific and granular the buckets, the closer the predictions are for a particular individual. But often the buckets are very broad and far from the needed granularity. They make strong assumptions about the items within each of the buckets that are in opposition to what makes us human—our uniqueness. Such systems serve to diminish this uniqueness by assuming that we share a large swathe of characteristics with the other members of the bucket. People whose needs are at the margins of the normal distribution are highly unique. Decision systems often lack enough data about people who are marginalized to form a digital footprint and thus these individuals get lumped into other buckets that developers and the machine learning system deem closest to them. This creates a negative, vicious cycle where their experience with the system is subpar, demotivating their participation and use of such a system thus further reducing their representation in these systems which entrenches the poor performance of the system.

Many labels, tags, or classifiers for the buckets are inherited from classification systems that were created prior to more progressive ideas about human rights. They reflect a discriminatory hierarchy and stereotypes. Many metadata ontologies are based on systems such as the Dewey decimal system. Dewey was known to be racist, sexist, and homophobic. These discriminatory attitudes find their way into the systems of classification.

This false homogenization of minority populations [17] and biased labeling are in part artifacts of the lack of diversity in the developer teams that build these systems and also in a lack of interdisciplinary work that could inform and highlight such potential pitfalls. While there is a move to bring together researchers from the humanities and technical sciences, the efforts fall short in involving people from the local communities that are closest to the problems they face and have the most in-depth understanding of the intricacies of how to address them. Empowerment of these communities to build and participate in the development of machine learning systems is urgently needed. This will also require a democratization of data science tools that are interoperable with alternative access systems.

5.6 Bias in the Training and Validation Phase

When designing machine learning systems, during the training and validation phase, it is a common practice to utilize a metric to tune the model to output the kind of predictions that are within acceptable limits as deemed by the developers of the system. Often the focus is on a single metric that might be combined from an ensemble of underlying metrics such as the F1 score which is the harmonic means of the precision and recall that gives the accuracy of the test being performed or the area under the receiver operating characteristic (ROC) curve which highlights the diagnostic ability of a classifier as a function of its discrimination threshold. While such metrics definitely help guide the development process, they present a challenge when it comes to evaluation of the performance of the model on any minority populations that will interact with the model. The reason poor performance can become obfuscated for minority groups in single metrics is that they form a small part of the overall training corpus and hence even when the error rates are high on this small subpopulation, the effects on the total are minimal because of the low error rates on the majority (which is what the model tuning roughly aims for). Thus, there is a need to have granular metrics that address as many as many subpopulations that will interact with the system as possible. The incidental benefit of such a design consideration is that they also put front and center a deep discussion on who the possible audiences are going to be and possibly an interdisciplinary consultation that helps to gain a humanities-oriented understanding to complement the technical considerations.

While single metrics can plague machine learning systems from a technical perspective, single metrics from a business perspective also create unintentional consequences. Take the case of YouTube Kids, a supposedly friendly alternative with content that is safe for kids to watch as opposed to the full gamut of content that is available on the larger YouTube platform. In late 2017 [18], it was discovered that there was an emergent problem with a proliferation of disturbing and macabre child-themed content that started hijacking the underlying algorithm for which content gets shown to users. With the focus of the platform on the metric of time spent viewing videos and how much ad revenue can be generated by the video views, it was natural that the system could be gamed such that videos with a high chance of being clicked and watched would be pushed up in the feeds of users. While initially the videos were manually generated by low budget studios that knew what the right combination of content and keywords was to get their videos to the top, they later automated the process further exacerbating the problem. The problem became so grievous and a PR nightmare that YouTube took active steps to take down as much of the offending content as possible. Another prime example for how news feed algorithms can be hacked and gamed is Facebook [19] which is widely known and understood.

5.7 Self-Reinforcing Bias

Discriminatory data systems can be self-reinforcing, amplifying existing measurement biases. One example is psychometric tests based on data gathered from one group, used to test another group. Another example is success metrics determined with one group, inappropriately applied to measure the success of another group in another context. This is exemplified in the assertion that data has been used to crack the "social impact genome," thereby holding the formula for the success of any social impact venture [20]. This assertion ignores the diversity and variability of social issues.

5.8 Transparency and Auditability

In mission critical contexts, where machine learning systems are used to make significant decisions about human lives, one of the biggest concerns is that they are inscrutable to human understanding because of representations [21] that are high dimensional and unintelligible in the normal human sense. The systems are thus also inscrutable to existing judicial and societal systems that protect human rights. Unless the systems are scrutinizable we risk being uninformed subjects to systems that are all powerful in making important decisions about how we live our lives.

This brings us to the subject of transparency and auditing of machine learning systems. Transparency and auditing comes in two forms: one in the form of analyzing a system by domain and technical experts, often in the employ of those that have developed and sold those systems; the other is a more public form of transparency where there is a published description that accurately reflects how the system works and people who are data subjects of such a system can gain a reasonable understanding and check the account against their experience with the system. The first form of transparency and auditing is weaker in protecting the public because of the limited scope of who can understand the system. This approach excludes those that might be the most vulnerable to the negative decisions rendered by the system.

This unnecessarily places the burden of proof on the vulnerable population. The second alternative provides greater access to protection for the general public. Unfortunately, this second alternative is often difficult to achieve because of intellectual property concerns and the difficulty of articulating the description of the system so it is broadly understood. This is one of the difficulties faced by the GDPR in demanding the right to explanation. There is some promise in terms of creating such descriptions modeled on datasheets for datasets [22], model cards for model reporting [23] and nutrition labels for datasets [24].

Audit of such systems in part relies not only on transparency but also on interpretability of the system. Audit is important because when legal claims need to be made by those that have been negatively affected by the system, there needs to be a demonstrable way of proving that the wronged party has been the subject of an injustice. This can include, but is not limited to, a trail of training data, model configuration, random seeds, states of machines, and any other artifacts of the system that help in reproducing the results that are in dispute. Without such reproducibility, legal recourse is limited, and justice is compromised as a result of the probabilistic nature of the systems.

5.9 Privacy and Protection Against Data Abuse and Misuse

Differential privacy [25] mathematically guarantees that seeing the result of a differentially private analysis makes the same inference about an individual's private information, whether or not that individual's private information is included as an input to a machine learning system. It stands in contrast to ad hoc anonymization where entities apply unverified mathematical constructs to obfuscate individual data in the hope of preventing reidentification. The key issue lies in the unverified portion of how these techniques are applied, essentially meaning that they can't provide any guarantees that they can avoid future attacks on the privacy of the data.

Several techniques such as the mosaic effect have in the past effectively demonstrated that ad hoc anonymization is grossly insufficient in addressing the issue of anonymization and protection of data privacy. The mosaic effect [26] refers to the combination of multiple datasets to glean information that wouldn't readily be accessible from any of the datasets on their own. There are several notable examples of this: Netflix had in 2004 released a dataset of anonymized movie ratings in a competition setting to elicit from the data science community better recommendation engine models. A couple of ingenious data scientists from Princeton were able to combine this data with publicly available information on movie ratings from the IMDB and utilize the rare and esoteric movies and their ratings with the data from Netflix and reidentify individuals. This led to the public disclosure of the sexual orientation of someone who hadn't come out of the closet to her family and coworkers. This gives us an idea of how the more unique an individual is, as is often the case for disabled individuals, the higher the risk they face when it comes to loss of data privacy in the digital realm.

Another example was the release of search data from AOL pre-2000 which when combined with publicly available information led to the disclosure of a great deal of personal information about a woman in mid-Western US including what medications she took, that she had a dog, etc. There are many other examples but these two serve to illustrate the gap between what ad hoc anonymization purports to do and the protections it can realistically provide. Differential privacy on the other hand provides mathematical guarantees within certain bounds on the amount of personal information that can be leaked from the use of a dataset. Most importantly, it is safe against future attacks even if new datasets are made public and can prevent the use of the mosaic effect to reidentify individuals. This is a very important property because data brokers utilize disparate sources of data, combining them to create rich profiles on individuals essentially identifying and targeting them for various purposes, most often for targeted advertising.

A caveat exists in the utilization of differential privacy as a tool where the stronger requirement for protection of privacy comes at the cost of accuracy in the subsequent analysis of the data. However, this is tunable by the epsilon parameter which provides granularity appropriate for the kind of data analysis being done and its requirements in terms of accuracy. Differential privacy is relatively new and lacks widespread knowledge. For the first time in the US Census history, differential privacy is being utilized [27] as a tool to meet data disclosure protection requirements as mandated by law. This will bring this technique into wider usage, and also serve as a public test of the efficacy of a mathematically grounded technique in data protection.

One way to reduce data abuse is to limit the unnecessary distribution of personal data. A notorious instrument of data overreach that unnecessarily causes the collection of unprotected personal data is the ubiquitous all-or-nothing terms of service agreement that requests permission to collect and use personal data in exchange for a service. These terms of service agreements are usually couched in legal fine print that most people could not decode even if they had the time to read them. Additionally, it has been noted that for all the services that someone uses throughout the year, it would take several weeks to review all of the user agreements. This means that it has become a convention to simply click "I agree" without attending to the terms and the rights that have been relinquished.

The Inclusive Design Research Centre (IDRC) has proposed a personal data preference standard as an instrument for regulators to restore self-determination regarding personal data. The proposed standard will be part of an existing standard developed by the IDRC and an international working group. The parent standard is called AccessForAll or ISO/IEC 24751. The structure of the parent standard enables matching of consumer needs and preferences with resource or service functionality. It provides a common language for describing what you need or prefer in machine-readable terms and a means for service providers or producers to describe the functions their products and services offer. This allows platforms to match diverse unmet consumer needs with the closest product or service offering. Layered on top of the standard are utilities that help consumers explore, discover, and refine their understanding of their needs and preferences, for a given context and a given goal. The personal data preference part of this standard will let consumers declare what personal data they are willing to release to whom, for what purpose, what length of time and under what conditions. Services that wish to use the data would declare what data is essential for providing the service and what data is optional. This will enable a platform to support the negotiation of more reasonable terms of service. The data requirements declarations by the service provider would be transparent and auditable. The standard will be augmented with utilities that inform and guide consumers regarding the risks and implications of preference choices. Regulators in Canada and Europe plan to point to this standard when it is completed. This will hopefully wrest back some semblance of self-determination of the use of personal data.

5.10 Opportunities to Address Bias

Optimistically, there are emerging opportunities to address existing bias in AI and machine learning. Given all the negative press and outrage that AI systems receive when they are used in situations where they have impacts on human lives such as credit scoring [28] and bail and parole decisions [29], there are many opportunities that arise where we can utilize these systems to battle against entrenched historical and stereotypical biases and injustices. For example, AI can be used as an effective tool to reveal and highlight these biases while providing people with the opportunity to potentially escape algorithmic determinism [30]. Simply stated when algorithmic systems are used to render decisions for people, they don't provide transparency regarding the factors used to make the decisions, e.g., in denying them a loan based on a series of factors including but not limited to the credit score of the individual. In the cases where there is a lack of transparency, which is often the case with systems using machine learning, the individual is not given the opportunity to improve their score or an explanation as to how they might be able to do better in the eyes of the system in the future. This creates a vicious cycle where the most disadvantaged are at the mercy of the decisions of such opaque systems. At the same time the people who utilize these systems to provide services also have their hands tied. They are not able to provide any sort of aid to the applicants because they lack understanding of how the system is designed and operated.

The primary cause of opaqueness (the so-called black box [31] nature of the systems), is not limited to the fact that there are barriers to inspecting them, but more that the patterns within the systems are not human interpretable. A crude analogy would be to think about what an apple means to a human brain when it sees one. The things that come to mind are its taste, texture, size, color, etc., but to a computer-vision system, an apple is merely a distribution of pixel densities with different colors. This representation presented to a human will not mean much, but the system will be able to identify the apple to a reasonable degree.

This inscrutable nature of AI is what we seek to address when we talk about explainable systems in the context of digital privacy regulations such as the GDPR [32]. Currently there is a legal debate regarding the way these requirements are phrased in the regulations. The crux of this debate is that explanations require an audience and specification of the level of technical and social understanding and knowledge such that the explanations are tailored to the audience. Lacking such a specification it is hard to know whether such a requirement has been satisfied and it consequently weakens any legal claims that disadvantaged parties can make. This may result in an accountability loophole for service providers. When it comes to technical measures, there are efforts underway to provide tools to the designers and developers of these systems to achieve a certain degree of explainability to their users. Some examples of these techniques include LIME [33], Alibi [34], SHAP [35], ELI5 [36], and others. The corresponding opportunity on the legal side will emerge from an interdisciplinary approach of lawyers who are versed in the technical aspects to further guide the developments of these tools and clarify the laws and regulations to meaningfully address issues of transparency and explainability.

Another pertinent opportunity is the creation of a consumer-facing tool that allows users of various systems to assess how their data is used in making decisions about them and how they might be negatively impacted because of inherent biases and lack of representation in the systems. The key challenge will lie in discerning what data is actually used in various services and how exactly they are used. Companies providing services might not be forthcoming in sharing all the details, citing intellectual property concerns.

One of the primary ways to address this is to involve people with expertise in under-represented needs, in the process, from the start. This includes the people and communities with lived experience of the marginalized needs. Secondly, one can also consult a nascent but growing literature [37] on building inclusively designed AI systems that cross the disciplinary boundaries of technology and social science. Unfortunately, most people with experience of the challenges also face many barriers to participation that predate the broad adoption of data analytics. Also, cross-disciplinary boundaries are often entrenched and strongly defended and social science is seen as secondary or nonessential in the race to dominate in AI.

A meta-conversation can be had here in terms of the voices in the AI ethics research community where a dominant minority through social media influence dominate and dictate agendas, funding and resources in the field, often at the cost of missing out on hearing from those that are closer to the problems of exclusion and bias and have faced them firsthand.

5.10.1 Advantages of Diversification

Creating AI systems and associated practices that leverage diversity, rather than sacrificing diversity for the sake of efficiency, helps in addressing some of the current shortcomings of AI for all users. One example is the detection of unwanted shifts. Distributional shift in the data, also known as covariate shift [38] is an important cause of concern in the deployment of machine learning systems, especially those that are online learning systems. The key reason being that there arises a discrepancy between the distribution of data that the machine learning models are trained, validated and tested on, versus the distribution of data that it encounters in the real-world setting. These gaps are to be expected but the problem emerges when proactive approaches are not taken to account for such gaps and instrumentation is not put in place to manage and adjust the models in response to the shifts. Another possible issue is the interaction of the system with out-of-distribution data (OOD), which is often the case with individuals with special needs as they are normally not well represented or not at all represented in the training datasets. Better representation in the dataset is one way to potentially mitigate the effects of the OOD and distributional shift, another more potent way is to set up guardrails to monitor prediction outputs, from the machine learning system in deployment, that can alert system maintainers of predictions running out of acceptable thresholds that can be set up according to the requirements of the task that the system is being used for. To achieve this, when systems are purchased from third parties, one must put in place a team that monitors the deployed system in a DevOps-like fashion rather than using the system out of the box without any modifications. Indeed, there is an emerging field called MLOps [39] that aims to bring some of these DevOps practices into the world of machine learning. Research has shown that deep ensemble models (which average predictions across several models) are more robust than vanilla approaches when it comes to distributional shifts and OOD data. This harnessing of several diverse models may be a promising direction for future research. Even small ensemble models have been shown to be effective thus not imposing a high computational burden.

5.10.2 Stepping-stone Principle

Neuroevolution borrows from the idea of biological evolution to design intelligence, or intelligent systems. The fundamental premise of biological evolution is that there isn't an overarching goal, yet it leads to interesting characteristics in the populations and species that emerge as a part of the process. Without a specific goal in mind, biological evolution evolved several different anatomical forms. One example is mechanisms for flight, in bats using segments of skin between the arms and torso and in the case of birds with hollow bones and feathers. Thus, without a specific, predetermined goal, flying emerged as a way to occupy a novel niche in variable ways. This is referred to as the stepping-stone principle [40] and it has shown success in designing novel architectures for intelligence. It relies heavily on novelty and emphasizes exploration over trying to optimize for a particular goal. By being open ended, the solutions that emerge are often better than the ones that are finely tuned to achieve a specific goal from the outset.

There have been numerous successes utilizing this approach, e.g., to beat a game called Montezuma's Revenge [41] where prior deep learning based approaches failed, this novelty-driven approach led to a high score that outperformed not only prior automated attempts but also human high scores. DeepMind has also expressed interest in this approach and their system to beat the game of StarCraft [42] has been quite effective at competing against highly ranked professional players.

This presents a potentially interesting angle to attack negative consequences of current AI systems, including ethics, safety, and inclusivity issues. When focusing too narrowly on specific goals, e.g., in credit allocation, we tend to come up with systems that penalize applicants that aren't within the norm. The phenomenon occurs with systems that are designed to serve the majority and ignore minorities with unique needs such as people with disabilities. Perhaps, creating systems via the stepping-stone principle can yield configurations that bypass such shortcomings in hitherto unforeseen ways. This may offer a better alternative to trying to achieve accessible and fair designs by working from a foundation with unstated and unchallenged assumptions. This novelty and exploration weighted approach would provide a fresh take on addressing problems and although computationally expensive, it may create a net positive externality by producing fair, just, and equitable AI systems.

An additional benefit that might arise from this would be that it can be paired with domain expertise as a light hand guiding along the evolution and selection of models that exhibit interesting characteristics. An alternate evolutionary pathway to creating AI systems that are more ethical, safe, and inclusive via this approach could then inform more traditional approaches. An open-ended model will stand to gain from an unbiased, *tabula rasa* means to achieve a goal without the trappings and stereotypes that get embedded from training approaches in the world of machine learning.

5.10.3 Move from Deep Learning to more Bottom-Up Systems

While traditional deep learning relies on large corpora of training data and approaches prediction from a top-down perspective, this view misses the finer details that arise in the case of the margins of the decision domain. To make meaningful predictions in such cases requires a higher degree of granularity, something that can be achieved using a more bottom-up approach utilizing small data [43] that has the added benefit of achieving privacy preservation by pushing model training and inference generation closer to the source of the data, often on device.

Additionally, when it comes to minorities (e.g., people with disabilities) the uniqueness of each individual (often as a minority of one) means that training in context, capturing the finer details that lie outside the statistical norm, makes for more meaningful predictions. This edge-computing, on-device systemic model also helps circumvent the reidentification problem that arises in the case of highly unique individuals. Folding in such individuals and making AI systems accessible and useful sets a great precedent for future scenarios which might have continued to be excluded otherwise.

This serves as a breeding ground for innovation that can help build more ethical, safe, private, and inclusive AI systems for all, while pushing technical boundaries in research and development to come up with techniques such as small data, federated learning [44], differential privacy [45], homomorphic encryption [46], and other novel strategies.

5.10.4 Removing the Advantage of Being the Same as Most People

An early experiment and challenge to popular machine learning methods is a manipulation of the traditional multivariate Gaussian curve, dubbed the "Lawnmower of Justice." Originally intended as a critical art form, the "Lawnmower of Justice" restricts the repeat of any data element within a dataset to a specific threshold, thereby cutting off or flattening the hill of the Gaussian curve [47]. This is in line with leveling the playing field of the data elements. This causes the machine learning engine to attend to the full spread of data and attend to weak signals. It takes longer to reach useful inference but is better able to address variability.

5.10.5 Distributed AI

Inclusive design is about the equilibrium achieved by including the broadest diversity within an integrated and evolving complex adaptive system. In this chapter we have explored the bias against diversity and variability within current AI systems. We have not addressed the inclusion or integration aspects or the social aspects of inclusive design. Inclusively designed AI that addresses needs at the margins requires systems that can use small datasets, are able to handle nonlinear, multivariate learning spaces; are organized bottom-up; are nonparametric or more exploratory and descriptive, rather than prescriptive; and, have emergent self-repairing properties without centralized supervision.

One domain of AI that holds promise is distributed artificial intelligence (DAI), including multiagent systems, distributed problem solving, and swarm intelligence. This is a topic beyond this chapter. The unintended consequences and failures of deep machine learning have made manifest the need to attend to the principles of inclusive design. DAI must synthesize the coordination of multiple agents, and holds promise for manifesting the social aspects of inclusive design.

5.11 Conclusion

AI has failed to achieve its promise and even caused significant harm. This harm is unevenly distributed but ultimately affects the system as a whole. This harm is most evident where inclusive design is not considered. The three dimensions of inclusive design are (1) recognize that everyone is different and attempt to achieve one-size-fits-one experience in an integrated system that supports self-knowledge and agency regarding personal difference; (2) ensure that the process of design is inclusive of the individuals that will feel the impact of the design; include as many diverse perspectives as possible; and (3) attempt to achieve benefit for all while cognizant of the complex adaptive system that is the context of the [48]. Current AI systems revert to an average, denying the full range of human difference; do not support or encourage participatory design; and, reduce rather than address complexity while optimizing for the majority or largest customer base. Inclusive design offers a promising strategy to address flaws in AI and fundamentally advance ethical AI innovation.

References

[1] https://www.globalpolicy.org/home/271-general/53036-data-is-the-new-gold.html.

[2] The Economist (2017). "The world's most valuable resource is no longer oil, but data". May 6th, The Economist: New York, NY, USA.

[3] Jacobs, S. (ND). "Fueling the Creation of New Electronic Curbcuts". The Center for an Accessible Society. http://www.accessiblesociety.org/topics/technology/eleccurbcut.htm

[4] Veale, M. and Binns, R. (2017). "Fairer Machine Learning in the Real World: Mitigating Discrimination Without Collecting Sensitive Data". Big Data & Society, 4(2), https://doi.org/10.1177/2053951717743530.

[5] Emerson, R. W. (2015). "Convenience Sampling, Random Sampling, and Snowball Sampling: How Does Sampling Affect the Validity of Research?" Journal of Visual Impairment & Blindness, 109(2), 164-168.

[6] https://qz.com/india/1133527/aadhaar-indias-intrusive-biometric-id-is-forcing-hiv-patients-to-forgo-treatment/.

[7] Raju, R. S., Singh, S. and Khatter, K. (2017). "Aadhaar Card: Challenges and Impact on Digital Transformation". arXiv preprint arXiv:1708.05117.

[8] https://www.bbc.com/news/world-asia-india-43207964.

[9] https://www.livemint.com/Politics/the2brDHqCcWztKlpLLeuK/Aadhaar-authentication-failure-doesnt-mean-denial-of-benefi.html.

[10] https://www.creativespirits.info/aboriginalculture/economy/internet-access-in-aboriginal-communities.

[11] https://www.forbes.com/sites/blakemorgan/2018/07/16/how-amazon-has-re-organized-around-artificial-intelligence-and-machine-learning/.

[12] https://www.wired.co.uk/article/how-do-netflixs-algorithms-work-machine-learning-helps-to-predict-what-viewers-will-like.

[13] https://benanne.github.io/2014/08/05/spotify-cnns.html.

[14] https://thenextweb.com/insider/2015/07/10/you-can-now-download-a-dataset-of-1-65-billion-reddit-comments-beware-the-redditor-ai/.

[15] Deng, J., Dong, W., Socher, R., Li, L. J., Li, K. and Fei-Fei, L. (2009). Imagenet: A Large-scale Hierarchical Image Database. In 2009 IEEE Conference on Computer Vision and Pattern Recognition, (pp. 248-255). IEEE.

[16] Simonite, T. (2018). "When it comes to gorillas, google photos remains blind". Wired, January, 13.

[17] http://news.mit.edu/2018/study-finds-gender-skin-type-bias-artificial-intelligence-systems-0212.

[18] https://link.medium.com/mJBdE4vO33.

[19] Lapowsky, I. (2018). "How Russian Facebook Ads Divided and Targeted US Voters Before the 2016 Election". Wired, April, 16.

[20] https://www.impactgenome.org/about-us/.

[21] Olah, C., Satyanarayan, A., Johnson, I., Carter, S., Schubert, L., Ye, K. and Mordvintsev, A. (2018). "The building Blocks of Interpretability". Distill, 3(3), e10.

[22] Gebru, T., Morgenstern, J., Vecchione, B., Vaughan, J. W., Wallach, H., Daumeé III, H. and Crawford, K. (2018). "Datasheets for Datasets". arXiv preprint arXiv:1803.09010.

[23] Mitchell, M., Wu, S., Zaldivar, A., Barnes, P., Vasserman, L., Hutchinson, B. and Gebru, T. (2019). "Model Cards for Model Reporting. In Proceedings of the Conference on Fairness, Accountability, and Transparency (pp. 220-229).

[24] Holland, S., Hosny, A., Newman, S., Joseph, J. and Chmielinski, K. (2018). "The Dataset Nutrition Label: A Framework to Drive Higher Data Quality Standards". arXiv preprint arXiv:1805.03677.

[25] Dwork, C. (2008). "Differential Privacy: A Survey of Results. In International Conference on Theory and Applications of Models of Computation (pp. 1-19). Springer, Berlin, Heidelberg.

[26] https://wws.princeton.edu/news-and-events/news/item/mosaic-effect-paints-vivid-pictures-tech-users-lives-felten-tells-privacy.

[27] https://www.census.gov/content/dam/Census/newsroom/press-kits/2019/jsm/presentation-deploying-differential-privacy-for-the-2020-census-of-pop-and-housing.pdf.

[28] https://qz.com/1276781/algorithms-are-making-the-same-mistakes-assessing-credit-scores-that-humans-did-a-century-ago/.

[29] Angwin, J. and Larson, J. (2016). "Bias in Criminal Risk Scores is Mathematically Inevitable, Researchers Say. Propublica, available at: https://goo.gl/S3Gwcn.

[30] https://www.oii.ox.ac.uk/blog/algorithmic-determinism-and-the-limits-of-artificial-intelligence/.

[31] Rudin, C. (2019). "Stop Explaining Black Box Machine Learning Models for High Stakes Decisions and Use Interpretable Models Instead'. Nature Machine Intelligence, 1(5), 206-215.

[32] Wachter, S., Mittelstadt, B. and Floridi, L. (2017). "Transparent, Explainable, and Accountable AI for Robotics. Science Robotics, 2(6), eaan6080, 31 May 2017. DOI:10.1126/scirobotics.aan6080

[33] Mishra, S., Sturm, B. L. and Dixon, S. (2017). "Local Interpretable Model-Agnostic Explanations for Music Content Analysis". In ISMIR (pp. 537-543).

[34] https://github.com/SeldonIO/alibi.

[35] Lundberg, S. M. and Lee, S. I. (2017). "A Unified Approach to Interpreting Model Predictions". In Advances in Neural Information Processing Systems (pp. 4765-4774).

[36] https://eli5.readthedocs.io/en/latest/overview.html.

[37] https://cyber.harvard.edu/ethics-and-governance-ai-reading-list.

[38] Snoek, J., Ovadia, Y., Fertig, E., Lakshminarayanan, B., Nowozin, S., Sculley, D. and Nado, Z. (2019). "Can you trust your model's uncertainty? Evaluating predictive uncertainty under dataset shift". In Advances in Neural Information Processing Systems (pp. 13969-13980).

[39] https://mlops-systems.github.io/.

[40] https://www.quantamagazine.org/computers-evolve-a-new-path-toward-human-intelligence-20191106/.

[41] Ecoffet, A., Huizinga, J., Lehman, J., Stanley, K. O. and Clune, J. (2019). "Go-explore: A New Approach for Hard-exploration Problems. arXiv preprint arXiv:1901.10995.

[42] Vinyals, O., Babuschkin, I., Czarnecki, W. M., Mathieu, M., Dudzik, A., Chung, J. and Oh, J. (2019). "Grandmaster Level in StarCraft II Using Multi-agent Reinforcement Learning". Nature, 575(7782), 350-354.

[43] Bagdasaryan, E., Berlstein, G., Waterman, J., Birrell, E., Foster, N., Schneider, F. B. and Estrin, D. (2019). "Ancile: Enhancing Privacy for Ubiquitous Computing with Use-Based Privacy". In Proceedings of the 18th ACM Workshop on Privacy in the Electronic Society (pp. 111-124).

[44] McMahan, B. and Ramage, D. (2017). "Federated Learning: Collaborative Machine Learning Without Centralized Training Data". Google Research Blog, 3.

[45] https://privacytools.seas.harvard.edu/differential-privacy.

[46] https://www.wired.com/2014/11/hacker-lexicon-homomorphic-encryption/.

[47] https://medium.com/datadriveninvestor/sidewalk-toronto-and-why-s marter-is-not-better-b233058d01c8

[48] https://medium.com/fwd50/the-three-dimensions-of-inclusive-design-part-one-103cad1ffdc2

6

Working with Big Data and AI: Toward Balanced and Responsible Working Practices

Valerie Frissen

eLaw – Center for Law and Digital Technologies, Leiden University, and SIDN Fund, The Netherlands

Abstract

Artificial intelligence (AI) has a huge potential to facilitate, enhance, and transform human activities, but concerns have arisen about the risks involved. A strong call for new ethical and regulatory frameworks has emerged helping us to build human-centered, responsible approaches to the use of AI. In this chapter, two cases of using ethical guidelines related to AI in working practices are described in detail. The first case addresses approaches to develop responsible AI in health care, more specifically for intensive care, whereas the second case focuses on using AI for data-driven approaches in the domain of public safety and organized crime. It is observed that in both cases particularly the issue of data quality causes problems, and a conclusion is that rigorous and standardized protocols will be required for collecting and using data for AI applications. Furthermore, the two cases demonstrate interesting differences and practical complexities in the way moral and ethical considerations are being taken into account.

Keywords: Artificial intelligence, ethical frameworks, responsible innovation, Big Data, working practices

6.1 Introduction

Whereas for decades the huge opportunities of digital innovation for our economy and society have been stressed, the current debate on the digital future seems to be colored more and more by concerns and fears. This is particularly the case in the field of artificial intelligence (AI) and autonomous systems (AS). In the domain of work and production for instance, the foreseen robotization of labor and production processes has resulted in serious concerns about mass unemployment and about the impact of technology on human agency and autonomy (see, for instance, Refs. [1, 2]).

AI has a huge potential not only to facilitate and enhance, but also to substantially transform human activities. On the one hand the further development of AI may boost the rise of new and innovative enterprises and may result in promising new services and products in—for instance—transportation, health care, education, and the home environment. On the other hand, it may also radically disrupt the way public and private organizations currently work and the way we live our everyday lives. Early excitement about the benefits of these systems has begun to be tempered and overshadowed by concerns about their risks. Actual concerns in this regard are lack of algorithmic fairness (leading to discriminatory decisions); manipulation of users; the creation of "filter bubbles" and disinformation; the infringement of privacy and safety and cybersecurity risks. And last but not least, concerns over possible abuse of the dominant market position of a few big players in the field [3].

In this context there is a strong call for new regulatory and ethical frameworks that help us to build a human-centric, responsible approach toward digital innovation. This approach is regularly framed as a new and specifically European perspective, distinctive from both the market driven Silicon Valley approach and the state driven approach toward AI in authoritarian countries such as Russia and China [4]. There is a growing consensus that a distinctive European approach toward AI should be human-centric, applying ethics and security by design principles and supporting the core values of the European Union: human dignity, freedom, democracy, equality, rule of law, and human rights [5]. The European Commission is taking the development of a "responsible" approach to AI quite seriously. Following its communication on an European approach to AI, published in April 2018, the Commission set up a high-level expert group on AI, consisting of 52 independent experts representing academia, industry, and civil society. In April 2019, the HLEG published the "Ethics Guidelines for Trustworthy

AI" [6]. This was immediately followed by a communication [7] in which the Commission fully adopted these guidelines and presented next steps, aiming at the implementation of the ethical guidelines in practice. The guidelines are based on fundamental rights and ethical principles and list seven key requirements that AI systems should meet to be trustworthy:

1. Human agency and oversight,
2. Technical robustness and safety,
3. Privacy and data governance,
4. Transparency,
5. Diversity, nondiscrimination and fairness,
6. Societal and environmental well-being, and
7. Accountability.

The HLEG acknowledges that these overall requirements are still quite abstract. To make them more usable in concrete working practices, the expert group has developed an *assessment list* that offers guidance on each requirement's practical implementation. Moreover, the HLEG stresses the importance of *piloting* these requirements in concrete practices and invites stakeholders to participate in this process. In this chapter this invitation is taken up. The chapter describes how ethical guidelines for AI are currently being operationalized in concrete working practices and explores the challenges and dilemmas this involves[1].

In this chapter, two cases are described in more detail. One case is a project called "AI will see you now," a research project carried out by the AMC Medical Research in a Dutch hospital[2]. This project aims to develop requirements for responsible AI in health care, more specifically for intensive care (IC). The second case is a project in which several stakeholders in the domain of public safety in the Netherlands are cooperating to develop a predictive data-driven approach for tackling organized crime, by using new methods of data analytics. The analysis of both cases is based on (research) reports, publications, and other project documents and on two in-depth interviews with the project leaders. Conceptually, the analysis of the cases builds on insights from philosophy/ethics of technology, particularly on

---

[1]Many lists of ethical requirements and principles concerning AI are currently being developed with a substantial amount of overlap. An overview of the current academic debate can for instance be found in Ref. [8]. A useful overview of codes and requirements can be found at https://www.rathenau.nl/en/digital-society/overview-ethics-codes-and-principles-ai.

[2]This project was funded by the Dutch SIDN fund after a call for proposals on "Responsible AI". The author is a director of this fund (https://www.sidnfonds.nl/excerpt).

an approach called "guidance ethics": This perspective sees technology and humans as inextricably interwoven[3] and not as separated from or opposed to each other. Ethically this implies that we need to develop a perspective that enables us to "live and work with" technologies in a balanced way, instead of drawing sharp moral boundaries between humans and technology.

6.2 Doing Ethics: Toward an Actionable Approach of Responsible AI

Although it is widely acknowledged that AI has a huge potential to facilitate and enhance our living and working practices, it is also clear that AI raises fundamental societal and ethical questions. In our view, an ethical perspective on AI involves (a) a thorough understanding of the potential *impact* of AI on society and on our everyday life; (b) a clear idea of what role we want to play AI in society, involving questions such as: what are the *values* to guide automation and automated decision-making; (c) the translation of these insights into new *value-driven methods and ethical/regulatory frameworks* for engineers, companies, and policymakers and, last but not the least, (d) "*machine ethics.*" Machine ethics is concerned with the question if it is possible to design artificial agents and algorithms in such a way that they behave morally, in accordance with human values.

Current developments in machine learning move toward stronger decision-making power and autonomy of machines. Greater autonomy comes with greater responsibility, and therefore we need a sophisticated system of checks and balances in the ecosystem in which AI plays a role. When more and more actions and decisions are delegated to technology—or are at the least more and more out of sight of human control mechanisms—this involves greater responsibility of all actors, including the technology itself. We need workable ways of explaining and contesting algorithmic decision-making, and ways of organizing human responsibility and oversight. Taken together, these developments call for a perspective which is now commonly referred to as "Responsible AI"[4]. The global and European community that is working on Responsible AI and is developing a common agenda for this field is

---

[3]E.g., Ref. [9]. A practical translation of these ideas can be found in Ref. [10] (in Dutch).

[4]This section is partly based on an unpublished advisory paper on Responsible AI for DG Connect, written by the author and professor Natali Helberger in their role as members of the CAF (Connect Advisory Forum, see https://ec.europa.eu/digital-single-market/en/caf-members). See also Ref. [11].

growing. Researchers, policymakers, and industrial and societal stakeholders recognize the need for approaches that ensure a safe, beneficial, and fair use of AI technologies, to consider the implications of ethically and legally relevant decision-making by machines, and to consider the ethical and legal status of AI itself[5]. Within the academic community researchers from distinct fields such as computer sciences, social sciences, law, philosophy, and ethics are finding each other. One of the prominent academics in this field, Virginia Dignum ([12], also [13], [14]), has described the new mind set in the AI field as follows:

> *Responsible AI is more than the ticking of some ethical boxes or the development of some AI add-on features in AI systems. Rather, responsibility is fundamental to intelligence and no system can be truly intelligent if it cannot understand responsibility [12: 4].*

The development of AI has for a long time been mainly driven by goals such as increasing performance and efficiency. Putting *human values* in the core of AI systems, calls for a mind shift of researchers and developers. They have to get used to putting goals such as transparency, explainability, accountability, and contestability first. A strong awareness among researchers and developers is needed of the potential societal impact of AI, as a first step to take responsibility for the systems that they are developing. This may translate into design and engineering principles, into codes of conduct (e.g., being accountable and transparent) and into new frameworks for regulation, legislation, and governance (e.g., liability). A still quite contested question is of a more fundamental nature: does this also involve *machine ethics,* i.e., mechanisms that enable AI systems themselves to reason about and act according to ethics and human values, and to justify decisions according to their effects on these values? Machine ethics may become particularly important when AI will continue to develop into self-learning, autonomous systems. This raise challenging questions such as: do we really want technology to act as a moral agent? And following from that, how can we define, measure, model, and optimize moral values? And is it even possible to define *shared* moral values?

---

[5]See for instance publications of the Institute of Electrical and Electronic Engineers (IEEE), such as their work on "Ethically aligned design," https://ethicsinaction.ieee.org and "Responsible AI. A Global Policy Framework," by The International Technology Law Association (2019).

For example, how do you teach a machine to algorithmically maximize fairness or to overcome racial and gender biases in its training data? A machine cannot be taught what is fair unless the engineers designing the AI system have a precise conception of what fairness is (..). Machines cannot be assumed to be inherently capable of behaving morally. Humans must teach them what morality is, how it can be measured and optimized. For AI engineers, this may seem like a daunting task. After all, defining moral values is a challenge mankind has struggled with throughout its history. If we can't agree on what makes a moral human, how can we design moral robots?[6]

An influential opponent of granting intelligent systems moral agency is Joanna Bryson, who argues—in a paper with the provocative title *Robots should be slaves*—that robots should neither be described as persons or human-like companions, nor should they be given legal or moral responsibility for their actions [15]. According to Bryson, robots are produced, programmed, and fully owned by humans. It is us, humans, who determine their goals and behavior, either directly or indirectly through specifying their intelligence or how their intelligence is acquired. Robots should be viewed as *tools* we use to extend or enhance our human abilities. In that sense robots "should be slaves." Although it is tempting to attribute humanoid qualities such as moral reasoning to machines, from Bryson's perspective moral responsibility ultimately has to remain within the domain of human actors. She thus makes a strict separation between machines and humans. From an ethical perspective only humans can and should be seen as moral agents, bearing responsibility for "what things can do."

A quite different position is taken by philosophers of technology who advocate a *mediation* perspective on the human–technology relation, such as Peter–Paul Verbeek[7]. These philosophers argue that a purely instrumental approach to technology—such as the one Bryson proposes—which sees technology as a mere *tool* in the hands of humans, fails to address the moral significance of technology itself. It leads to a humanist bias in ethics that ignores the actively mediating role of technologies. From a mediation perspective, it is argued that technology mediates between us and the world

---

[6]https://medium.com/@drpolonski/can-we-teach-morality-to-machines-three-perspectives-on-ethics-for-artificial-intelligence-64fe479e25d3

[7]This work is based on post-phenomenological thinking in the Philosophy of Technology. See for instance [16], [17]. See also [18] and [19].

we live in. Technologies help to shape human experiences, actions, and decisions and therefore they actively contribute to the way we live our lives. As they shape the way we experience and give meaning to our world, they fundamentally also have *moral* qualities. The invention and application of ultrasound technology, for instance, not only enabled us to "see" unborn children but it also induced new moral questions about (the ending of) unborn life. Face recognition technology is a technology that—be it intentional or unintentional—infringes human rights such as the right to anonymity and privacy. Our definition of privacy changes because of these technologies which in turn lead to new ways of "living with" and regulating these technologies. Algorithms are oftentimes based on systematic and repeatable errors, for instance in the way data are coded, collected, selected, or used to train the algorithm, leading to a reproduction of bias. This may lead to unfair decisions which systematically privilege one group over others. All these examples show that technology is never neutral; technological artifacts are both reflecting and shaping our values and behavior.

> *The mediating role of technologies can have a distinctively moral dimension. By helping to shape our practices and the interpretations on the basis of which we make decisions, technology can play an explicit and active role in our moral actions [17: 12).*

The mediation perspective advocated by Verbeek does not place humans and technologies at each other's opposite, but sees them as inextricably interwoven. The ethical implication of this viewpoint is that technologies should not be seen as "invasive powers in need of ethical limits" [17: 153], boundaries that can only be drawn by human actors. Technologies are an intrinsic part of our human existence and therefore what we need is an ethical approach that helps us to "live and work with" these technologies. Oftentimes ethical approaches toward technology take a humanist, and in that sense "human in control" perspective toward technology. This often results in an externalist and judging position, or, to use a metaphor, ethics functions in this perspective as a "border guard": defining the boundaries of "good and evil." A mediation perspective on the other hand looks at the ways humans are actually interacting with technology and how they mutually shape each other in the process. The aim of an ethical approach here is not to set limits or to draw boundaries, but to build a constructive relationship between people and technology. Again using a metaphor, ethics in this perspective can be seen as a "couples therapist": it helps humans to build a meaningful and responsible relationship with technology. Verbeek refers to this approach as "accompanying technology" (see the last chapter of Refs. [9, 10, and 20]), a concept he borrowed from the Belgian philosopher Gilbert Hottois [21].

This more hybrid way of looking at the human/technology relation may serve as a starting point for developing an actionable framework for responsible AI. Following from this approach is that moral agency and ethical responsibility are *distributed among human and nonhuman actors in AI*. This involves both the morality of the technology itself and of the many stakeholders involved in technology development, from researchers and designers to policymakers, industry actors, and end users. The relationship between humans and technology is shaped in *every* stage of technological development. A responsible approach toward AI thus cannot be limited to an ethical committee that defines the "boundaries," of what is allowed or not allowed; neither can it be restricted to responsible design principles only. It is fundamentally a *distributed and shared responsibility* of all stakeholders and actors, including the technology itself. This perspective also stresses the importance of empirical research of concrete technologies in their specific context of use and in terms of their specific social and ethical implications, whereas other ethical approaches would look at the human/technology relation in a more abstract and generalized way. The approach favored by Verbeek is thus both descriptive and normative: it is a combination of a concrete *descriptive analysis* of a technology in its specific context of use, and *doing ethics*: identifying points of moral reflection, assesment, and intervention in the process of technology development.

Recently, in a Dutch publication Verbeek and others presented a practical approach on how to translate this ethical approach into concrete action strategies [10]. A distinction is made between three basic options for "doing ethics": ethics by design, ethics in context, and ethics by user. *Ethics by design* focuses on doing ethics in the design stage of technology. No technology is neutral, it was argued above: every technology has built-in values which induce and shape certain behavior. This also implies that technologies can be *designed* in such a way that they express values that are considered important in a specific context (e.g., privacy by design or by default principles; internet filters that block certain unwanted content, internet standards that are based on distributed power principles, etc.). *Ethics in context* focuses on doing ethics in a broader organizational or societal context. This involves developing social norms and rules to protect and sustain the values that we deem important, such as codes of conduct, principles of good governance, regulation, and laws (e.g., the GDPR [General Data Protection

Regulation]). And finally, *ethics in use* focuses on awareness of key values when we use or apply specific technologies and the concrete responsible behavior this results in (e.g., explaining how automated decision-making systems have had an impact on citizens).

This "accompanying" or "guidance" approach to ethics seems to have resonated in the Responsible AI community as well. Virginia Dignum has made a more or less comparable distinction between different ways of applying ethics in AI practices [14:2]: (a) ethics *by* design: the technical/algorithmic integration of ethical reasoning capabilities as part of the behavior of artificial autonomous system; (b) ethics *in* design: the regulatory and engineering methods that support the analysis and evaluation of the ethical implications of AI systems as these integrate or replace traditional social structures; and (c) ethics *for* design: the codes of conduct, standards, and certification processes that ensure the integrity of developers and users as they research, design, construct, employ, and manage AI systems.

In the next two paragraphs the two cases introduced earlier will be presented. I will first describe the case, and the challenges and dilemmas that have come up when developing and using AI. Next, I will focus on how *actionable* approaches for responsible AI have been developed in these particular contexts.

6.3 Case 1: The Use of AI in Intensive Care

The first case is a project carried out by a multidisciplinary team of experts in a Dutch academic hospital with the title "AI will see you now" [22]. The objective of this project was to explore the concerns and needs of both physicians and patients regarding the use of AI to support decision-making in IC. It is well known that IC is a medical specialty dedicated to treatment of patients with (acute) life- threatening conditions. Therefore, often drastic decisions have to be taken, for instance, on the continuation of life-sustaining therapies. The costs of treatment in an IC department are high, both economically and emotionally. Economic IC costs in the Netherlands are estimated at around 20% of the total hospital budget. Evidently, emotional costs of IC treatment are also very high, and patients and their relatives not necessarily prefer intensive, life-sustaining treatments. Being able to make a reliable estimate of the prognosis for survival with an acceptable quality of life, in accordance with the preferences of the patient, is therefore essential for optimal patient care.

Although the use of algorithms in health care is not new, until recently both the amount and quality of data used as input for these algorithms were not sufficient from an AI perspective. Because of the growing use of Electronic Hospital records (EHR), electronic patient records (EPR), and consumer health apps, the amount of health data has grown exponentially in recent years. The use of AI to generate knowledge from these data is now more and more seen as a promising strategy to keep health care safe, accessible, and affordable. In this particular project an algorithm that can predict the chances of survival of IC patients was used as a case example, the so-called prognostic algorithm. The doctors interviewed in the project expressed a clear need for supporting information that could help them to estimate the chance of survival more objectively. Moreover, it was expected that prognostic information could help IC patients and their relatives to participate in the decision-making process in a better, more informed way. However, the use of AI in such life-or-death matters also raises many ethical questions. Questions in this project were among others: can AI indeed help all involved stakeholders to make better (informed) decisions? What would be the impact on trust in patient–doctor relationships? What does this mean in terms of accountability? Who is ultimately responsible for decisions being taken? The objective was to translate these questions and considerations into a set of specific guidelines for the use of algorithms in (IC) health care.

According to the interviewed project leader, IC doctors underscore the necessity of more objective, evidence-based decision-making processes and in that sense they welcome AI applications that may help them to make sound decisions. "It is the human factor that makes health care human, but at the same time that is also its weakness," was a comment given in the project. Research has shown, for instance, that ICU physicians' estimations of prognoses are modest and frequently overestimated. A treatment once it is initiated is often continued, although the reasons for doing so are not necessarily based on objective considerations (the so-called "sunken cost effect"). For doctors to use these prognostic algorithms, it is crucial that they are *easy to use* (preferably a simple "button in the "EPR") and that they are always *explainable*. Explainability of AI is valued very high by them: doctors must always be able to explain themselves in simple terms how and why a certain decision is taken[8]. This is considered crucial for trust in the doctor–patient relationship. Similarly, for the patients and their

---

[8]So not necessarily explainability of the algorithms themselves, but of the decisions that have been made by using these algorithms.

relatives the *transparency* of the decision-making process is very important and following from that, the communication between doctors and patients and the explainability of the role of AI in this process. However, patients also stress that communication is important in any case, and doctors quite often "tend to be a black box themselves." Doctors and patients further agree that there always needs to be a good balance between the human and the technological factor, or as one of them expresses this, a combination of "gut feeling with a strong support of data." In this project the potential benefits of the use of AI for the process of decision-making were thus clearly acknowledged by both patients and medical professionals, but human agency and oversight in any case were considered to be crucial as well. The human "line of arguing" in decision-making processes should be explicit and traceable, and ultimately the doctor's or the medical team's decision should always be leading. In terms of the high-level expert group's list of requirements, besides *human agency and oversight*, particularly *transparency and explainability* were thus valued very high by the stakeholders involved in the project.

One of the key *challenges* lined out in this project was that for the development of responsible AI, the amount of data no longer seems to be the problem, but the *quality of data* still is. Particularly in the IC, vast amounts of data are being collected through the continuous use of many high-tech monitoring technologies. Besides, the use of EHRs is now a given fact in Dutch hospitals, which implies that almost all aspects of medical treatment can now be digitally documented and datafied. From an ethical perspective, it is, however, extremely important that the quality of data being used as input for prognostic algorithms is very high. Life-or-death decisions can never be taken on the basis of inaccurate or biased input. This project shows that there is still a lot to be improved here, particularly concerning the EHR data. At this point in time, in EHRs data are not yet collected in a rigorous and accurate way and are often based on subjective decisions and interpretations. The question is of course whether this subjectivity is a bad thing or something that can be (and even should be) avoided. The problem is not so much the subjectivity per se, but the fact that subjective decisions and interpretations are not well documented and made explicit. Data used for decision-making may thus lead to a false sense of objectivity and security. Not only EHR data, but even the "hard" and structured data collected by the high-tech machines in the ER, are somewhat flawed, as the *decisions behind* these systematic measurements are regularly not well documented either. Moreover, data may not be complete: what is for instance often missing in EHRs is data about pre- and post-IC quality of life factors. For predicting survival, data about the quality of life of the patient, and what may happen after the patient leaves the hospital, are very important. And finally, there are vast differences between data collected by different hospitals, which makes them quite difficult to compare. The case thus makes it quite clear that, before taking steps toward using prognostic algorithms, we need to be very certain about the data that will serve as input for these applications. Quality of data is a major challenge. For that reason, the project leader concludes that it maybe wise for now to use only simple models, models that allow for a large amount of control. Her expectation is that because of this the use of machine learning, particularly in complex decisions such as whether or not to continue life-sustaining IC treatments, is still far away.

The next question, then, is how to cope with these challenges. And can the approach lined out above that made a distinction between three concrete strategies for acting (ethics by design, ethics in context, and ethics by user) be useful here? Based on the results of the project a set of 22 requirements has been found by the research team, clustered into 6 thematic groups[9]. These requirements can be seen as a starting point for "doing AI ethics in the IC."

6.3.1 Ethics by Design

As described in section 2, technologies must be designed in such a way that they express the values considered to be important in a specific context. This can be translated into both "machine ethics" and into various forms of ethical reflection in the design process. The description above has shown that in this particular case transparency, explainability, human agency and oversight, and quality of data are considered to be key values. This has been translated into design requirements for both the data used as input for the algorithms, and the algorithms themselves. According to the researcher, good examples of how control of *quality of data* can be guaranteed are quality of care registries such as the Dutch National Intensive Care Evaluation registry (NICE). For a registry such as the NICE, it is of great importance to acquire correct and unambiguous information in a privacy preserving way. To do so, among other

---

[9]These six clusters are the following: Requirements regarding which medical decisions can be supported with prospective AI models; Requirements regarding the quality of (medical) data that are used to build these models; Requirements regarding the developers of these models; Requirements regarding the development and the evaluation of these models; Requirements regarding the use of these models in the clinical practice; and Requirements regarding the communication about the predictions made by these models.

specific data requirements, rigorous quality checks and privacy preserving procedures are put in place *before* the data collected is used for any type of analysis. It is required, when developing AI algorithms by using data from the EHRs that their example should be followed. Furthermore, human reasoning must be reflected in the data, e.g., why something was or was not done. Also needed are more rigid and standardized protocols for documenting treatment decisions in the EHRs. A useful step to be taken here maybe to automate quality of data controls to a certain degree. And finally, a clear distinction needs to be made between data that can be used for research and for clinical purposes, the latter clearly asking for more rigid requirements than the first. For the design of *algorithms,* it is required that the code is always open and verifiable. Besides, algorithms used in clinical practices should always be freely available, meaning that their potential commercial value could never lead to restrictions of use for health-care purposes. Furthermore, algorithms need to be thoroughly substantiated scientifically. They need to be transparent in terms of chosen population, data source, possible bias in data, variables used, preprocessing, weighting, correction, and model types. The outcomes of the use of an algorithm must always be explainable (at least on a level of the model's global behavior), and simplicity is in that sense very important (preferably a "simple button in the EHR," as the doctors requested).

6.3.2 Ethics in Context

This specific dimension of doing ethics focuses on the broader organizational or societal context. It involves developing social norms and rules, such as codes of conduct, principles of good governance, regulation, and laws (e.g., the GDPR). In terms of regulation, particularly *privacy regulation, IP, and liability regulation* are considered important. Although compliance to privacy regulations is now quite obvious, in terms of privacy *governance* there is still work to be done. A step to be taken is the development of *impact assessments and risk analyses* as a requirement for using AI in the medical practice, as well as the monitoring and evaluation of these impact and risk assessments. Feedback loops and insight into errors or unforeseen consequences is considered to be very important here. Furthermore, the need for setting up *multidisciplinary* teams is stressed in this project: ideally a mix of knowledge of medical data and the medical context, knowledge of methodology, ethics, communication, and medical psychology. And evidently, patients (interests) should be well represented.

6.3.3 Ethics in Use

This dimension focuses on both the awareness of key values when applying AI, and concrete responsible behavior in everyday working practices. This maybe the most difficult dimension of doing ethics, as it requires a certain degree of internalization of ethical behavior. For instance, the quality of data aspect that is considered very important in this case, requires a high standard of "data hygiene" of all stakeholders involved. And although this may partly be automated or be translated into regulation and codes of conduct, the key is that all actors experience this as a shared responsibility. In many cases, this will require behavioral change. "Good behavior" maybe stimulated by investing in education and the improvement of both data skills and ethical skills and overall "algorithmic literacy" among all stakeholders.

6.4 Case 2: The Use of Data Analysis and AI to Prevent Undermining Crime

The second case is a project called "Citydeal: Data driven insights on organized crime" (hereafter called the Citydeal project), in which a large number of stakeholders in the domain of public safety in the Netherlands[10] is cooperating to develop a predictive data-driven approach for preventing and tackling "undermining crime." The concept of undermining crime refers to the targeted use by organized crime of institutions, practices, and opportunities in the "upper world." By using existing societal structures for criminal purposes, the normal functioning of society is undermined. The intertwining of the under- and upper world manifests itself in for instance infiltration by criminal networks in (local) politics and administration, the growth of shadow economies, and the facilitation of crime in the upper world by actors such as financial advisors, notaries, technicians, real estate agents, or the transport sector [23, 24, 25]. The Citydeal project is quite innovative because of the large number of participants who have expressed an intention

---

[10]This so-called "Citydeal" is a cooperation between the cities of Almere, Amsterdam, Arnhem, Breda, Den Haag, Eindhoven, Groningen, Helmond, 's-Hertogenbosch, Maastricht, Rotterdam, Tilburg, Utrecht, and Zwolle. Other partners are the Ministries of Internal Affairs & Kingdom Relations, Finance, and Justice & Security; the National Tax Service, the Netherlands Public Prosecution Service, and the National Police and Statistics Netherlands (CBS). More info (in Dutch) can be found at https://www.zichtopondermijning.nl. The midterm report, covering Phase 1, (2017–2019), can be found at https://www.cbs.nl/nl-nl/maatwerk/2019/44/city-deal-zicht-op-ondermijning-1e-fase

to act as "one government." This cooperation enables cross cutting data analysis on a national scale and is expected to provide new insights into the nature and scope, manifestations, and local roots of organized crime. New is also the close cooperation between public safety experts and data analysts from different bodies of the government on the one hand and data specialists of Statistics Netherlands (CBS) on the other. CBS is the national statistical office, which provides reliable statistical information and data to produce insight into social issues[11].

The objectives of the Citydeal project are twofold. First of all, it aims to contribute to a preventive approach to undermining crime. It is expected to enable the identification of (local and regional) patterns, opportunity structures, and vulnerabilities in specific sectors and branches. These patterns may than be translated into indicators for more effective policy strategies that help to address crime prevention and law enforcement. Second, the project aims to develop a better insight into the possibilities and limitations of advanced methods of data analysis to be used for such a preventive approach. It addresses questions such as what is the necessary quality of data, what are the practical possibilities and limitations of data analysis, and what are the legal constraints to take into account? The official goals of this Citydeal project are described in a covenant, which was published in the Staatscourant[12] in August 2017 [26]. The two themes that were chosen for in-depth analysis are property fraud and drug crime.

6.4.1 Approach

In the midterm report presented to the Dutch parliament in October 2019[13], the approach of the project is described in more detail. It mainly uses quantitative methods; techniques to objectify and quantify criminological

---

[11]CBS was established in 1899 in response to the need for independent and reliable information that advances the understanding of social issues. This is still the main role of the CBS. CBS performs public service tasks but operates independently and not under the direct authority of a Dutch ministry. Its independent position is enshrined in a specific law: The Statistics Netherlands Act (CBS wet). This law constitutes the legal basis for the CBS. Moreover, it operates strictly within the framework of several other legal frameworks concerning data (such as the GDPR).

[12]De Staatscourant is the Dutch "government gazette": the periodical publication of the Dutch government that has been authorized to publish public or legal notices. Since July 1, 2009, the paper is published online as "officielebekendmakingen.nl" (official announcements).

[13]https://www.tweedekamer.nl/kamerstukken/detail?id=2019D42854&did=2019D42854

phenomena. It mostly uses CBS data, which has two major advantages: the national scale on which research can be conducted, and the guaranteed reliability and confidentiality of the data. More precisely, it uses CBS *microdata*: linkable data (at the personal, company, and address level) which—in compliance with the CBS law—may only be used under strict conditions for statistical research. For this case this implies that the data can (a) never be used for law enforcement purposes, (only for research) and (b) can never be traced to individual addresses, persons, or companies. Data are pseudonymized which means that identifying personal characteristics are replaced by a unique code. In the project both descriptive statistics and data science techniques are applied, such as predictive models. The research is partly based on hypotheses and partly has an exploratory character, in the sense that it looks for unexpected patterns in the data that may lead to new hypotheses and predictions. All data selected are analyzed within the secure research environment and within the regulatory framework that applies to the CBS. The embedding in the CBS context is an important design principle for the project, because the strict CBS regulation can be seen as a formalized way of safeguarding not only the quality of data and methods, but also the protection of public values such as privacy, fairness, and transparency. Requirements for the use of data in such a way that these public values are not jeopardized, are laid out in the CBS law.

Both in the midterm report, the related documents, and in the interview with the project leader it is clear that a responsible approach toward this new form of data analysis is valued highly in the project. The use of predictive methods of data analysis cannot be done, according to the project leader, without being aware of the risk that important public values are under pressure. Therefore, the protection of *privacy*, the prohibition of discrimination (*fairness*), the *reliability of data*, and methods of analysis and maximizing *transparency* are explicitly taken into account. In the interview the project leader explained that an earlier project on address fraud has led to parliamentary questions that could not be answered satisfactorily. Therefore, the minister who is responsible for this spin-off project has demanded maximum transparency in the Citydeal project, implying that every step in the project needs to be logged and documented, to provide a sound and transparent basis for political decisions that maybe taken because of this project.

In the midterm report it is acknowledged that existing regulatory frameworks (such as the CBS law, police law, and the GDPR) broadly define the normative context for acting responsibly, but nevertheless these legal requirements need to be operationalized in more practical terms. From a more practical perspective, it is quite likely that dilemmas will occur in concrete working practices. The data analysis may, for instance, lead to risk profiles that are quite interesting for investigating and law enforcement purposes. Although the profiles themselves maybe used for law enforcement purposes, the underlying data may not: the use of profiles cannot in any way lead to potential *individual* suspects. The project leader notes that the law enforcement partners in the project, such as the police, really regret this: the use of the underlying data could be quite helpful in their everyday policing practice. This shows a struggle between conflicting public values: protecting public security through efficient and predictive investigation and law enforcement practices on the one hand, and safeguarding privacy and fairness/nondiscrimination on the other. The more concrete results of the project may shed more light on such experienced dilemmas and give a more detailed idea on how a responsible approach is operationalized in the project.

6.4.2 Results

The first goal of the Citydeal project was to contribute to a predictive approach to undermining crime by identifying patterns in the data that point to opportunity structures and vulnerabilities in the upper world. The second goal was to gain more insight into the possibilities and limitations of the new methods of data analysis applied in the project. In terms of the usefulness of these methods for predictive data analysis, the conclusion in the midterm report is positive. The datasets used in the Citydeal have appeared to be quite useful to gain more detailed insights into underlying patterns of organized crime and how these undermine the structures of the "upper world." These insights are considered to be helpful for developing a preventive *policy* approach toward undermining crime. According to the project leader, the patterns found in the data are significantly different from regular patterns. They help the project partners to use scarce law enforcement resources in a more targeted and thus more efficient way. Furthermore, these patterns are useful to develop predictive approaches that may prevent specific target groups to get into trouble in the future.

The project focused on the themes of *property fraud* and *drug crime*. An example of a pattern found in the data related to property fraud, was that data about inexplainable home ownership and about convicted homeowners were not evenly distributed across neighborhoods, but often concentrated in so-called *hotspots*. As a next step, the midterm report suggests that the knowledge about these hotspots with a high-risk profile can be used to search more effectively for "transit homes" (houses that are being sold more than 5 times in a period of 20 years). Another example of a pattern that was found concerning drug-related crime was that in one city the age of people dealing in hard drugs was significantly lower than in the other participating cities. This insight can be helpful to develop a preventive local policy targeting a specific age group. A third finding was that suspicion of certain criminal offences in the past, such as money laundering or possession of hard drugs, may help to predict future drug-related suspicions. The project also looked into family networks related to drug-related crime. The midterm report states that the insights into the role of family relations can be used as input for a more targeted and personalized approach toward drug-related crimes[14]. Overall, the conclusion is that these kinds of insights can be very helpful to increase resilience to organized crime, because they enable a "theme- or neighborhood-oriented approach."

The suggestions for further actions mentioned in the report and by the project leader raise the question of how the boundaries between research and law enforcement are actually drawn in concrete working practices. It is obvious that in the project itself a clear-cut boundary has been drawn between what is allowed in the context of research and in the context of law enforcement. This is guaranteed by strictly playing within the rules set by the CBS, which do not allow to trace back findings to the individual, personal level ($N > 10^{15}$), to use them in such a way that this leads to discrimination or to use these data for law enforcement objectives. Nevertheless, it is clear that the *findings*—the patterns found in the data—are in themselves quite sensitive, particularly because they are based on microdata on a personal (e.g., household and family relations) and neighborhood level. It is allowed to use these findings (the profiles for instance) in *another* context (law enforcement) in a way that may have unwarranted effects on values such as privacy and fairness: think of predictive policing practices that are based on neighborhood stereotyping. Although the overall approach of the project is responsible in the sense that it explicitly takes into account these public values, and

---

[14]In Dutch: PGA ("Persoons Gerichte Aanpak").

[15]To minimize the risk of disclosure of individual cases, in accordance with the Statistics Netherlands Act, numbers are rounded to tens (10). In addition, results based on less than 10 observations are suppressed or, in the case of counts, rounded to 10. If combinations of insights from different parts of this dashboard lead to small groups, those groups are so small in size that no firm conclusions can be drawn from the figures.

has put into place certain safeguards to protect these values, that does not necessarily mean that it is also responsible in its *effects*. We have found no clear indications that the project has considered such an *ex post* responsibility. A way to approach this maybe to include impact assessment tools in the next steps of the approach and to come up with checks and balances to protect these values in the next steps after the research stage.

What is striking in this project is that responsibility is framed mostly and almost exclusively in <u>legal</u> terms. In the relevant documents (midterm report, Staatscourant, website, and factsheet) and in the interview, the CBS context with its specific legal requirements and constraints is underlined as the main guarantee for the protection of public values. The report does not shed further light on how this translates into more practical challenges or dilemmas and how they are dealt with in concrete cases. The risk of taking such a formal and "legalistic" perspective is that responsibility is not seen as a shared responsibility; as long as you act within the boundaries of the law, you do not have to think about what this means in the more messy reality of everyday life work practices.

6.4.3 Doing Ethics

In terms of the three strategies for *doing ethics* as described earlier (ethics by design, ethics in context, and ethics by user), we can conclude that the strict embedding in the regulatory framework of the CBS shows that in this case doing ethics is mainly reflected in *ethics by design and ethics in context*, but much less in *ethics by use* (the everyday practices of users). *Ethics by design* is applied in how CBS regulations are translated into design principles, such as the requirements for validation of scientific quality and transparency of data and algorithms (e.g., logging everything, no black boxing). Furthermore, analysis of data is by design on group level ($N>10$), and personal data are always pseudonymized (privacy by design). This is enshrined in the legal context that is taken as the starting point for this project. *Ethics by context* is in this case interpreted in a mostly legalistic way. The rules to adhere to are explicitly laid out in the covenant that was published in the Staatscourant. Furthermore, within the CBS context several rules and formalized practices apply, through cooperating with the CBS, also apply to this project (e.g., 4-eyes principle, output controls checking on privacy, and nondiscrimination). Finally, the rule that data may only be used for research purposes and not for law enforcement follows from the contextual embedding in CBS regulations as well. In terms of *ethics by user*, however, there is much less explicit evidence how this is envisioned in the Citydeal project. The legalistic perspective that is favored in the project evidently *prescribes* how data and data analysis methods must be applied in a responsible way. However, this does not necessarily entail a shared responsibility and understanding of what constitutes responsible behavior among all stakeholders involved.

6.5 Discussion

In its AI strategy—launched in 2018—the European Commission has made an explicit choice for a "human centric" approach toward AI. The Commission considers AI to be a strategic technology: it can bring great benefits to the European society and economy. At the same time, it is acknowledged that AI raises serious legal and ethical questions and may have a considerable impact on the future of work. A sound ethical and legal framework is deemed necessary for increasing trust in AI among European citizens. AI should not be seen as "an end in itself, but as a tool that has to serve people with the ultimate aim of increasing human well-being"[16]. Although the EU has relevant legal frameworks in place, such as the GDPR and the Cybersecurity Act to strengthen trust in the online world, AI brings specific challenges that cannot be fully addressed by existing regulation. Particularly the ability of AI to learn and make decisions without human intervention is potentially problematic. The Commission has therefore set up a high-level expert group on AI with the task to come up with a set of workable ethical guidelines. The expert group has identified seven key ethical requirements and an assessment list to help check whether these requirements are fulfilled. However, to make these "high over" and quite general guidelines really usable, the *specific context* in which they are applied should be taken into account. For this purpose, the Commission invites stakeholders to pilot the application of these requirements in concrete working contexts and to evaluate whether the guidelines require further adjustment.

   To give an example that illustrates the complexity of what we are dealing with here, the Commission (following the expert group's guidelines) states that AI should be "transparent" and "explainable," which is operationalized in the following way:

---

[16]https://ec.europa.eu/digital-single-market/en/artificial-intelligence

"It is important to log and document both the decisions made by the systems, as well as the entire process (including a description of data gathering and labelling, and a description of the algorithm used) that yielded the decisions. Linked to this, explainability of the algorithmic decision-making process, adapted to the persons involved, should be provided to the extent possible. Ongoing research to develop explainability mechanisms should be pursued. In addition, explanations of the degree to which an AI system influences and shapes the organizational decision-making process, design choices of the system, as well as the rationale for deploying it, should be available (hence ensuring not just data and system transparency, but also business model transparency). Finally, it is important to adequately communicate the AI system's capabilities and limitations to the different stakeholders involved in a manner appropriate to the use case at hand. Moreover, AI systems should be identifiable as such, ensuring that users know they are interacting with an AI system and which persons are responsible for it."[17] (COM (2019) 168 final, p.5)

It is not hard to imagine that a random company or governmental body that sincerely wants to use AI in its daily operations and has the intention to do this in a responsible way, is left with a slight attack of despair after reading this (and this is only one of the seven requirements).

In this chapter we have taken up the challenge to investigate how ethical guidelines such as the HLEG ones, are actually being applied in concrete working practices, by focusing in more detail on two cases. The first case is a project that develops requirements for responsible AI in health care, more specifically for IC. The second case is a project in which a predictive data-driven approach for tackling organized crime is being developed. Both cases show that a value driven and responsible perspective on AI is taken quite seriously and is in fact at the heart of both projects. A striking similarity in both projects is the issue of *"data quality"* that comes up as the issue that causes the biggest headaches. This underscores clearly, that any application of AI—before we even start talking about algorithms or responsible uses of AI—in fact has to begin with a thorough and meticulous quality assessment of the data that is being used for building and training algorithms. In both cases it is stressed that quite *rigorous and standardized protocols* for collecting

---

[17]https://ec.europa.eu/transparency/regdoc/rep/1/2019/EN/COM-2019-168-F1-EN-MAIN-PART-1.PDF

and using data need to be in place. For healthcare purposes—particularly in the IC—this is obviously very important: to develop algorithms that help physicians to make treatment decisions, one needs to be absolutely sure about the accuracy and validity of input data. In practice, this causes huge challenges, as vast amounts of data are needed for useful AI applications. Although in terms of quantity, the potential of EHR-, EPR-, and consumer health app-data is promising; however, in terms of quality there is still a lot to be improved. One of the practical ways of approaching this challenge in the current circumstances is to work with rather limited applications of AI, "simple models" that allow for a lot of human control. Another solution is to distinguish sharply between research purposes and clinical purposes: what maybe allowed (to a certain degree) in a research context can never be allowed in the clinical practice.

A similar approach can be observed in the organized crime case. In this case, data may only be used for research and policy development but *never* for law enforcement purposes. This follows from the choice made in the project to only use CBS data. CBS—or Statistics Netherlands, is strictly regulated by law, which also enshrines regulations about the use of data. By using the CBS data, some of the necessary rigid data quality checks (such as reliability) are covered. Quality of data in this specific case also implies that values such as privacy (e.g., pseudonymization of data, profiles may never be traced back to individuals), fairness (nondiscrimination), and transparency (everything has to be logged) are addressed as well.

Looking at the seven HLEG requirements, in both cases particularly *privacy, transparency, and explainability* are seen as important (ethical) values to be safeguarded. Moreover, in the IC case the *human agency and oversight* requirement is also stressed. In the IC practice, it is considered very important to strike the right balance here. Treatment decisions can obviously never rely completely on algorithmic decision making. In a hospital context, a "human in command" approach is strongly favored. AI is merely seen as a useful *tool* to help physicians to make better decisions, grounded in rich data. This is closely related to the transparency and explainability requirements. In the medical context this is mainly translated in terms of *communication*. In doctor–patient interactions, doctors always have to be able to explain the treatment decisions they make: in that sense explainability of algorithmic decision-making is seen as very important by both doctors and patients. If it is not possible to do this in an easy and simple way, this may well be the key barrier to the acceptance of AI by both doctors and patients. In the organized crime case, the value of fairness is addressed more explicitly. This is understandable in the context of the ongoing debate on predictive policing: biased data may lead to discriminatory profiling and policing practices.

Finally, the two cases show interesting differences in their approach of what we have called "doing ethics." Conceptually, we have outlined a "distributed ethics" approach in this chapter, inspired by the mediation perspective in philosophy of technology. In this approach moral agency and ethical responsibility are distributed among all actors and stakeholders in technological development. Not only humans are considered to be "moral actors," technology has moral qualities as well. In practical terms "moral agency" of technology implies that data, code, and algorithms are not neutral instruments, but value driven and value-loaded in their design. Moral and ethical considerations thus have to be taken into account from the early design stage of technological development until the stage of adoption and use. In the IC case, a set of requirements has been developed that can be seen as a multifaceted and indeed a "distributed ethics" approach, covering a mix of design principles, regulatory and organizational standards, protocols and codes of conduct, and the application of ethics in concrete practices of use (e.g., stimulating awareness and "algorithmic literacy" among stakeholders). In the organized crime case to a certain extent a similar multifaceted approach is visible. However, this approach is very strongly framed in a "legalistic" perspective. By complying to the strict regulations of the CBS law, most moral and ethical considerations are assumed to be covered. The risk of such a legalistic approach may be that ethics is not experienced as a shared and distributed responsibility that constantly needs to be reconsidered in concrete and dynamic real-life practices.

Both cases show that applying a responsible AI approach is not an easy thing to do and is definitely more complex than publishing a list of ethical principles on one's website. A distributed ethics approach does not necessarily make this task easier, but it does provide us with a perspective on how to develop a shared responsibility toward AI. The cases show that in each context, important values are weighted differently, different stakeholders have different and sometimes competing interests, and all this is translated into specific ways to act responsibly. Cases such as the ones described in this chapter may help to take ethics of AI out of the ivory tower of policy makers and ethical committees and make it a real shared responsibility.

References

[1] The Economist (2016). "Automation and Anxiety. Will smarter machines cause mass unemployment?" Available online: https://www.economist.com/special-report/2016/06/23/automation-and-anxiety.

[2] Pew Research Centre (2017). "Automation in Everyday Life". Available online: https://www.pewresearch.org/internet/2017/10/04/automation-in-everyday-life/

[3] Frissen, V., Lakemeyer, G. and Georgioupoulos, G. (2018). "Ethics and Artificial Intelligence". Available online: http://bruegel.org/2018/12/ethics-and-artificial-intelligence/.

[4] Van Dijck, J., Poell, T. and De Waal, M. (2018). "The Platform Society: Public Values in a Connective World". Oxford: Oxford University Press.

[5] Craglia, M. et al. (eds.) (2018). "Artificial Intelligence: A European Perspective". European Commission Joint Research Centre. Available online: https://publications.jrc.ec.europa.eu/repository/bitstream/JRC113826/ai-flagship-report-online.pdf

[6] High Level Group of Experts on AI (2019). Ethics Guidelines for Trustworthy AI. European Commission, Brussels. Available online: https://ec.europa.eu/futurium/en/ai-alliance-consultation/guidelines

[7] European Commission (2019). "Communication from the Commission to the European Parliament, the Council, the European Economic and Social Committee and the Committee of the Regions - Building Trust in Human Centric Artificial Intelligence", COM (2019), 168.

[8] Mittelstadt, B., Allo, P., Taddeo, M., Wachter, S. and Floridi, L. (2016). "The ethics of algorithms: Mapping the debate". Big Data & Society, July–December 2016, 1-21. https://doi.org/10.1177/

[9] Verbeek, P.P. (2011). "Moralizing Technology. Understanding and Designing the Morality of Things". Chicago: The University of Chicago Press.

[10] Verbeek, P.P. and Tijink, D. (2019). "An approach to 'guidance ethics': a dialogue about technology with an action perspective". Leidschendam: ECP | Platform for the Information Society. (In Dutch: Aanpak Begeleidingsethiek. Een dialoog over technologie met handelingsperpectief).

[11] The International Technology Law Association (2019). "Responsible AI. A Global Policy Framework". Available online: https://www. itechlaw.org/ResponsibleAI

[12] Dignum, V. (2017). "Responsible artificial intelligence: Designing AI for human values". ITU Journal: ICT Discoveries, Special Issue No. 1, Sept. 2017.

[13] Dignum, V. (2019). "Responsible Artificial Intelligence: How to Develop and use AI in a Responsible Way". Switzerland A.G.: Springer Nature.

[14] Dignum, V. (2018). "Ethics in artificial intelligence: Introduction to the special issue". Ethics and Information Technology, 20, 1-3.

[15] Bryson, J., (2010). "Robots should be slaves", In: Y. Wilks (ed.), Close Engagements with Artificial Companions: Key Social, Psychological, Ethical and Design Issue. John Benjamins Publishing Company, pp. 63-74.

[16] Verbeek, P.P. (2005). "What Things Do. Philosophical Reflections on Technology, Agency and Design". University Park, PA: Pennsylvania State University Press.

[17] Verbeek, P.P. (2011) "Moralizing Technology. Understanding and Designing the Morality of Things". University of Chicago Press.

[18] Winner, L. (1980). "Do artifacts have politics?" In: "Daedalus", Vol. 109, No. 1, Modern Technology: Problem or Opportunity? (Winter, 1980), pp. 121–136. The MIT Press.

[19] Ihde, D. (2014). "A phenomenology of technics", In: Scharff, R. and Dusek, V. (eds.), Philosophy of Technology. The Technological Condition. An Anthology. Second Edition. John Wiley & Sons, pp. 539-560.

[20] Verbeek, P.P. (2010). "Accompanying technology, philosophy of technology after the ethical turn". Techné: Research in Philosophy and Technology, 14(1), Winter, 49-54.

[21] Hottois, G. (1996). "Symbool en techniek: over de techno-wetenschappelijke mutatie in de westerse cultuur". Kok Agora.

[22] Klopotowska. J. (2019), *AI WILL SEE YOU NOW. Het perspectief van de arts en de patieĺnt op een verantwoorde toepassing van Artificiélle Intelligentie (AI) ter ondersteuning van medische beslissingen.* (The perspective of the doctor and the patient on a responsible application of Artificial Intelligence (AI) to support medical decisions). Available online: https://kik.amc.nl/KIK/reports/TR2020-01.pdf

[23] Tops, P. and Tromp, J. (2019). "De achterkant van Amsterdam; een verkenning van drugsgerelateerde criminaliteit". Amsterdam: Gemeente Amsterdam.

[24] Boutellier, H. et al. (2020) "Een einde aan ondermijning", In: Tijdschrift voor Veiligheid 1.

[25] Steen, M. van der, Schram, J., Chin-A-Fat, N. and Scherpenisse, J. (2016). "Ondermijning ondermijnd". Den Haag: Nederlandse School voor Openbaar Bestuur.

[26] Staatscourant, 2017, nr 48699, 29 August 2017.

7

Working with Digital Technologies: Complexity, Acceleration, and Paradoxical Effects*

Christian Korunka

Department of Work and Organisational Psychology, Faculty of Psychology, University of Vienna, Vienna, Austria

Abstract

Digital technologies affect our lives in many ways. This chapter gives an overview of the potential general effects of digital technologies on the quality of working life. First, a short overview of the general development of digital technologies in the world of work is presented. Focus is on the potential general effects of "conventional" digital technologies (i.e., computers, office technologies, and information and communications technology at work places) on quality of working life. Next, theoretical approaches explaining these effects are presented. Two scientific concepts are especially useful in explaining the effects of these technologies: (1) the social acceleration concept of Hartmut Rosa and (2) the concept of paradoxes in digital technologies. Based on these concepts, new demands related to digital technologies in the current world of work are discussed and examples of empirical studies investigating these new demands are presented. As flexible work is a general feature of the current world of work and usually strongly digitalized, its different forms and potential effects on quality of working life are identified. Finally, evidence-based recommendations with the aim

---

[1] An earlier version of this chapter was published in a German language book [1].

137

of increasing quality of working life when working with digital tools are introduced.

Keywords: Social acceleration, quality of working life, new job demands, need based job design

7.1 New Digital Work Demands

7.1.1 Dispersion of Digital Technologies in the World of Work

Technologies have always played an utmost significant role for human development in general and also for the development of work. Since the Industrial Revolution, which was peaking already more than 150 years ago, the importance of technological developments for work has ever been increasing. The evolution of innovations has often been a wave-like process, with phases of several decades [2]. Often mentioned examples of such development phases are the formation of railway transport and travel in the 1850s, the rapid technological developments in the "roaring twenties" (1920s), and particularly the "digital technologies revolution" since the 1970s. The rapid spread of computer technologies was the most consequential development in this context. The most well-known example is the development of the Internet. A forerunner of the Internet in the 1960s, although with a very limited distribution mainly in the scientific realm, was the ARPA net. From then on, it needed another 25 years for the development of the first web browser at the CERN institute in Geneva. The "world wide web"—as we know it today —was started 1991 and opened for the public in 1993. Since then an extremely rapid development took place. Figure 7.1 shows the development of the number of Internet users over the last 25 years.

In the early 1990s, the Internet was available mainly in the United States and Europe, and primarily used by scientists and at universities. In the developing countries, the spread of the Internet started about 10 years later. Currently there are about 90% of the North Americans, nearly 90% of the Europeans, 52% of the Asians, and about 40% of Africans using the Internet [3].

A strong increase in computer use is also found in the world of work. Figure 7.2 shows the development of computer use at work places in five selected European countries.

As shown in Figure 7.2, currently at more than 50% of the work places in the European Union a considerable part of the working time is spent with computers. Still, there are noticeable (but decreasing) differences between the

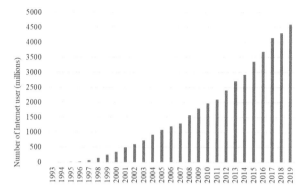

Figure 7.1 Development of the number of Internet users [3].

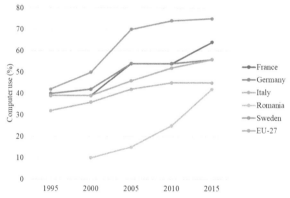

Figure 7.2 Development of computer use at work places in five selected European countries [4].

member countries observable. The former Warsaw Pact members catched up in their share of computer use over the last decades.

Digital technology use has many advantages and shows positive effects even on the quality of working life. However, there are also many negative outcomes found in the literature (for an overview, see Ref. [5]). The effects of digital technology use are found on many levels: individual, jobs, teams, organizations, families, and on the society in general. The following paragraphs sum up examples of these effects, focusing mainly on digital technologies currently used at work and at home (computer use, cell phone use, communication technologies, etc.)

Individual outcomes: Beside the obvious advantages of digital technologies use such as the extensive access to any kind of information, there maybe positive follow-on effects like an increased sense of independence and autonomy. Social media has at least the potential for a better integration in the community, for instance when people are able to stay in contact even over long distances, different time zones, etc. Many services maybe easily reached from any place and at any time.

However, there are many challenges for individuals, like the requirement to find a personal balance between easy and comprehensive information access and the meaningful use of that information. Digital technologies use has also a potential for negative effects on the individuals. For instance, permanent use of digital technologies at work or at home might be accompanied by rigid body postures, leading to carpal tunnel syndrome and musculoskeletal disorders. Such effects were already observed with the first introductions of computers at work places. "Techno strain" and "techno addiction" are new forms of negative outcomes of digital technology use [6]. Social media may lead both to a strong invasion of privacy and to dependencies on social networks, or even to social isolation [6].

Outcomes on the job level: Beside many improvements of work processes, there is a potential for many positive effects of digital technologies use on job design aspects. Digital technologies may for instance help to decrease physical workload, to improve feedback processes, and to support learning processes. On the other hand, learning new things maybe an opportunity, but it could easily change into a barrier or even a risk, when there is a permanent need to learn new things in ever shortening intervals [5]. Many digital technologies result in the shortening of response times, which at the same time leads to increased time pressure and intensification of work. Permanent connectivity at work and at home leads to permanent availability, which in turn increases job pressure and strain at home. Maybe one of the biggest challenges accompanied by digital technology use at work is the potential increase of work interruptions. Communication via e-mail or social media makes it very simple to interrupt other people during their work. Many researchers observe an increase in burnout over the last decades and are thus also referring to the increased use of digital technologies.

Outcomes at the organizational level: Cost reductions, economic advantages, and quality improvements are some of the more obvious goals of digital technology implementations in the organizations. Another important positive aspect is the possibility for collaborations and teamwork over long distances. However, there are also many challenges for organizations when implementing and using digital technologies. The number of implementation processes related to digital technologies is quite often very high and also leads to strong demands for the employees affected by these implementations. Digital technologies also have a potential to increase control on many levels. With the introduction of new digital technologies, there is always the risk for impairment of organizational trust. New digital technologies often lead to certain and increasing demands for leaders and managers and they permeate organizational levels and processes, leading to complex technological interdependencies throughout the organizations. The dependency on digital technologies maybe the biggest threat at the organizational level [5].

Outcomes at the work–family-interface: Many potential positive effects both, for telework and mobile workplaces, are described in the literature. Work and family life maybe better integrated, and commuting times maybe reduced. On the other hand, the potential improvements of telework and mobile work seem to come at a price. Many studies dealing with mobile work settings describe increased work–family conflicts and higher strain for many teleworkers [5].

Outcomes at the societal level: At the societal level, there is also a huge potential for both positive and negative outcomes when using digital technologies. For instance, technologies brought improvements for many "risk groups," such as elderly people, for handicapped people, and for persons living on the edge of the society. At the same time, digital technologies may also increase mutual dependencies at the societal level. Furthermore, the appearance of a "digital divide" may exclude certain population groups from the advantages of digital technologies.

New work contracts and gig economy: Beside the abovementioned changes on the individual, job, organization, work–family, and societal levels, there is a general trend observable leading to a change in contracts and a move toward a "gig economy" society. This development has many facets. A decrease of the classic long-term fixed contract (9/5; permanent contracts with job tenure) is accompanied by an increase of short-term contracts, hourly-based work, and general decreases in job security. Based on the opportunities of the Internet new, mostly low-level jobs emerge all over the world. There is a trend observable toward self-employment and entrepreneurial jobs. Furthermore, the increased development of robot technologies will change many jobs and make many jobs unnecessary. Experts are discordant in their expectations about the extent of such changes over the next decades, but many believe that there will be a reduction of work opportunities in many work areas. All these changes will affect us on all of the above-mentioned levels. In general, these developments will further contribute to the already existing "digital divide." Many employees will have better jobs and will work less, but other groups will have strongly reduced opportunities on the job markets.

This cursory enumeration of potential effects and outcomes of digital technologies not only shows the strong potential of digital technologies to change and influence our lives and the world of work, but also, that there are both positive and negative outcomes of these technologies (for more details, see Ref. [5]). All these effects may further be increased in the future by the rapid development of digital technologies and their increasing complexity. But what are the drivers of these technologies?

7.1.2 Globalization and Acceleration as Drivers

One of the most far-reaching changes in the world of work in the 20th century was the transition from a mainly production-based focus of work to new forms of service work. This massive change took place only within a few decades. During these times, the significance of material resources and means of production decreased, whereas knowledge gained increasing significance. At the same time a rigid division of work and standardized process execution was replaced by more flexible structures and processes.

The process of globalization also plays an important role in these developments. The rapid increase of the internationalization of markets leads to an increasing competition, even at the national level, as well as to increased competition pressures in all economic sectors. Because of deregulation processes ("neoliberalism"), an increased competition arises at all levels of society, even up to the level of national states, finally leading to a massive reduction of social benefits and social protections. These developments are further accelerated by global interconnectedness of individuals and organizations, contingent by the possibilities of modern digital technologies.

Hartmut Rosa [7] provided with his concept of "social acceleration" a widely accepted and comprehensive description pattern for such phenomena. The social acceleration concept allows the integration of many of the processes described here into a comprehensive societal frame model and especially a better understanding of the role of digital technologies in these developments. Rosa describes three interrelated aspects of social acceleration in late capitalistic societies:

Technological acceleration is the intentional and goal-directed acceleration of processes, like the increases of transportation speed, communication, and production. A prominent example of technical acceleration is the rapid increase of communication speed. Only a few years ago, letters were used for communication, which had a response time of some days. Nowadays, e-mail and social media enable real-time communication over arbitrary distances. Digital technologies are the core driver of technical acceleration. Moore's law [8], based on his observation that the power of digital circuits doubles every 12–24 months is maybe the most well-known example of the speed of technological acceleration (even if some authors claim a reduction of the speed of hardware development in the last decade).

Acceleration of social change is the increase of the decay rates of norms, expectations, and relationship patterns. An example is the increase of learning processes due to fast technological development, especially in the continuous implementation and use of new digital technologies. Further examples are the increasing rates of change processes within organizations, or the increase of changes in personal relationships, e.g., the changing numbers of "friends" in social networks.

Acceleration of the speed of life describes the continuous increase of action- and experience episodes per time unit. In the world of work, this relates both to objective changes, like the (measurable) increase of operation speed or the shortening of break times, and to subjective experiences, like the appraisal of increasing time shortages or increases in time pressure, but also the fear of not being able to cope with the speed of change processes. Technological acceleration is in many circumstances the main driver of the two latter aspects of acceleration.

A consequence of the acceleration of social change is the fact that many people in the work force experience that they are thrown back to themselves in many decisions because of missing requirements and targets, but also of the decrease of social safety nets. Employees need to take more responsibilities for their working conditions, their achievements of their objectives, and also for their social coverages such as health insurances, pension plans, etc. Pongratz and Voß [9] coined the term "labor force entrepreneur" to describe a new form of a working person, described by increased self-control and independent planning, steering, and control of their own work. This leads clearly to new demands for the employees in accelerated societies.

7.1.3 New Job Demands and Paradoxes

Thus, the three aspects of social acceleration are the drivers of new demands in the current world of work. In our own research, we have distinguished and investigated three new job demands as direct consequences of the acceleration of social change and the acceleration of the speed of life [10].

Work intensification describes a process characterized by an intensification of work effort. Specifically, work intensification captures the intensive effort that is needed to complete more tasks within 1 working day. In this sense, it can be distinguished from extensive effort or longer working hours (intensification). Extensive effort does not necessarily mean that work intensifies, because the extension of work does not imply that work itself intensifies; rather it could also arise from a long hours working culture. Thus, work intensification is defined as a multifaceted construct characterized by the need to work faster and face tighter deadlines, by a reduction of idle time, and the need to conduct a number of work tasks simultaneously [11].

Example: In an interview study we recorded a typical example of work intensification: "Last year we had a goal attainment in our sales department of 103%. This year, they make the 103% to 100%; if you could reach the goals ones, you could reach them always. Next year, we have again 103%, based, on the "100%" of this year. It always increases. I do not know how long we will be able to cope with it." (workgroup manager, telecommunication)

Intensified autonomy demands: Autonomy, usually defined as action- and decision autonomy, is one of the most important resources at work. The classic Karasek model [12] postulates that all jobs are characterized by two general features, namely job demands and job control. Whereas job demands refer usually to quantitative demands, like work load, time pressure, or— see above—work intensification, job control (or autonomy) refers to the discretion employees have over deciding how to pursue their tasks. Karasek [12] posits that only the absence of control/autonomy leads to negative effects on the quality of working life of employees, independent of their amount of workload. In a similar vein, the more recent job demands-resources model [13] postulates that job resources buffer the negative effects of job demands; autonomy is seen as one important job resource. Already a few decades ago it was argued that job autonomy, while clearly an important job resource, may lead at very high levels to negative effects on the quality of working life. For instance, Peter Warr, in his widely recognized "vitamin model" [14], postulates a curvilinear relationship between autonomy and quality of working life, with positive effects of autonomy only up to a certain level: very high levels of autonomy, then transfer into a "job necessity" instead of a resource, leading to inhibiting action rather than fostering it. The current flexibilization trends are also related to accelerated working conditions in the modern world of work, e.g., flexibility in time and location. This results in new demands especially at the boundaries between work and family. It is quite obvious that such very high autonomy conditions (like "boundaryless" work where employees are able to work anytime and anyplace) lead to the fact that these conditions are experienced as highly demanding. More and more employees need to actively plan where and when to work, which is not only demanding by itself, but also by the fact that social contacts with other colleagues or team members need to be planned. Furthermore, autonomy demands refer not only to the fact of planning where, when, and how to work. Even job careers need to be planned much more actively as in earlier times, when one job in many cases was a lifetime job, executed at similar time schedules (9/5) and at the same place over an entire career. Since autonomy is still an important resource, autonomy demands are usually experienced as highly demanding and highly motivating at the same time (challenge demands).

Example: An office secretary worked for many years in the same office, 5 days a week 9/5. Recently the company moved into a flexible office, accompanied by flexible working conditions. Now the secretary needs to decide on a daily or even hourly base not only what to work, but also where and when to work, and organize the work and especially the work communication adequately.

Intensified learning demands: Accelerated social change processes at work appear as continuous organizational change processes. Job profiles are changing and organizational units are restructured. Many employees experience an ever-increasing number of such processes in their organizations. New management concepts are implemented, departments are reorganized, and new hardware and software are permanently implemented in their work places. The speed of the introduction of new management trends is also increasing. All these processes lead to intensified learning demands for employees. Specialized knowledge, which is needed to execute the jobs, needs also be updated at an increasing pace. Similar to intensified autonomy demands, learning demands are also often experienced as highly demanding and motivating at the same time.

Example: The latest management fashion is "agile" management, an iterative, incremental method of managing activities in engineering, information technology, and other business areas that aim to provide new product or service development in a highly flexible and interactive manner. Such a management form results in an intensive learning process for everybody affected in an organization.

Thus, it is noticeable that both intensified autonomy demands and intensified learning demands often lead simultaneously to positive, i.e., increased motivation and job satisfaction and negative, i.e., job strain and burnout, outcomes for employees. This refers to the currently used digital technologies, but maybe even more to the technologies in the pipeline of development (robots, artificial intelligence [AI], etc.). This conclusion is also confirmed by research and reflects many personal experiences when using new digital technologies, which can be fascinating and draining at the same time. Therefore, many researchers talk about "paradox" effects when analyzing the effects of digital technologies on the quality of working life of employees. A paradox is defined as the simultaneous occurrence of contradictory elements, which are not only inseparably intertwined but occur also over prolonged periods in time. Discreetly viewed each element maybe considered as logical and expectable regarding its effects. However, the occurrence of different elements at the same time makes them inconsistent and irrational [15]. Related to digital technologies use, the following paradoxes are discussed in the research literature.

Autonomy paradox: Autonomy in the world of work was especially in the earlier history of work and organizational psychology always seen as an important and exclusively positive resource at work. The positive effects of autonomy were the cornerstone in many theoretical concepts and frequently confirmed by empirical research studies. In many job design approaches, the increase of autonomy of the workforce is seen as a central goal, not only to improve motivation and quality of working life, but also as a human goal itself (e.g., [16]). Autonomy is the basis for control of the execution of job tasks and enables free decisions. Already about two decades ago, some studies investigating highly autonomous teams found that with increasing autonomy the mutual control of the team members is also increasing (e.g., [17]). Other empirical studies confirm that employees having full autonomy in their schedules, working times and work places often intensify their work effort [18]. Thus, high autonomy seems to increase both self-control and the control of the work of others. The development of new norms and values of increased work effort in highly autonomous work groups, the need to maintain professional reputation and external deadlines (in some cases from other highly autonomous team members or working groups) affect employees directly in their high autonomy. At the same time, autonomy disguises the fact that these norms are forms of organizational control [19]. The autonomy paradox describes these immanent and complex relationships between autonomy and control, which could be carried out in the form of self-control or by mutual control of other team members. Especially evident is the autonomy paradox in job with high flexibility in time and space creating highly mobile working conditions where quite often self-driven work intensification, indicated for instance in "voluntary" prolonged working hours, is observable.

Example: A well-confirmed example of the autonomy paradox is the work outcome observed in telework settings. Most studies confirm that highly autonomous work settings (where employees work without any form of external control) are accompanied by an increased outcome on the side of the employees.

Connectivity paradox: In many jobs, the use of digital technologies supports work at anyplace and anytime. Digital technologies allow access not only to all data necessary for executing the tasks, but also to connect to coworkers by many communication channels (e-mail, social media, skype, etc.). Exactly this combination of fully autonomous work at the same time in permanent connection with flexible work, there arise the need to be connected with others—managers, customers, coworkers, etc. This need is perfectly supported by new forms of digital communication tools and platforms. Highly flexible work raises the expectation that these employees are easily reachable not only via digital technologies, but also at any time. As a consequence, persons working highly flexible and/or at home are no longer able to draw clear borders between work and private life. Out of the feasibility to actively design flexible working conditions and draw clear borders between work and nonwork results in a permanent connectedness, at anytime and anyplace. This could even lead to a full dependency from digital technologies in the form of digital technologies addiction (see [20, 21]). Another paradox effect related to the connectivity paradox is the fact that interruptions due to the permanent communication needs and communication attempts of others increase even though the flexible work settings (for instance in the home office) would have the potential to strongly reduce the interruptions and work fully undisturbed. This is highly relevant as interruptions are one of the most important stressors at work.

Example: Many of us (especially "knowledge workers") work at home. In a potentially undisturbed work setting we work on our studies and manuscripts. But there is always the possibility of opening another browser window to check the weather, check the news of fulfill a need for online shopping, and so on. Finally, the option of permanent connectivity leads to permanent interruptions on basically undisturbed work places.

Ter Hoeven et al. [22] describe the entangling of autonomy and control or flexibility and connectedness as the "practical paradox" of new technologies. Based on the job demands-resources model [23] they analyzed the complex interaction of demands and resources when working with new technologies explaining the simultaneous occurrence of strain reactions and high work engagement, and thus the appearance of positive and negative well-being when working with digital technologies.

7.1.4 Flexible Modes of Work

Digital technologies are not only drivers of social acceleration processes but they are also catalysts for the development of many forms of workplace flexibility over the last years. Workplace flexibility started already some decades ago when the conventional 9/5 working time arrangements were slowly replaced by many forms of flexible working time schedules. Recent empirical studies based on representative samples of the workforce (e.g., [24]) show a rapid development toward more flexibility in many aspects of jobs. Four forms of workplace flexibility can be distinguished:

Flexibility in time: There exists a wide range of flexible time concepts, ranging from flexible time schedules (usually with core times where employees need to work in their offices) to part-time work and trust based working hours. The latter concept is defined by a full dissolution of fixed working times. Work is then regulated only by the agreement and control of work targets and not by time schedules.

Flexibility in place: Telework settings were implemented in some organizations already decades ago, but most often only for smaller groups of employees. In telework, employees have a clearly defined second workplace, besides their office desk, usually in their homes. More recently and based on the possibilities of newer digital technologies applications, highly mobile workplaces have been introduced. With this form of work, there is no fixed workplace such as a main office. Work maybe performed at any place, and usually also at any time. Based on the recent ILO data [24], about 12% of the European workplaces could be described as highly mobile work. Most of these workplaces are some forms of office work, knowledge work, different forms of service work, and management. Recent data from Austria [25] point to a strong increase of this type of work.

Flexibility in work organization: Project work, which was a few decades ago carried out as narrowly defined projects, is a widely used form of work management today. Temporary projects have replaced the conventional line organization in many organizations. Other examples of flexible organizing are virtual teams as a form of collaboration in distributed and/or international organizations or different concepts of agile management.

Flexibility in work relations: Triggered by the economic crisis in 2008, many permanent work contracts with social coverages were replaced by temporal and/or short time work contracts. An extreme form of such a contract is "work on demand," where employees only work when they are really needed, on an hourly or daily basis. Many organizations, for example in German-speaking countries, moved from permanent to labor leasing contracts. Developments like the "gig" economy, where people all over the world work on a pay per piece base, without any securities are the most advanced form of such developments. While such contracts may have positive opportunities for some employees (e.g., in low-income countries), for many other people such form of contracts result only in high uncertainties. Distinct from the previous forms of work flexibilities, which usually lead to paradoxes and combines positive and negative outcomes for employees, flexibility in work relations has mainly negative effects on the quality of working life of employees affected by measures such as leasing contracts.

7.2 New Work Demands and Quality of Working Life

A secondary analysis of recent data from the European Working Conditions Survey [4] confirms the increase of deadline pressures between 1995 and 2010 in many European countries; although at the round of data collection (2015) these increases were somewhat mitigated (Figure 7.3).

The pronounced differences between the countries are seen in work intensification. This and other data show clearly that a predominant part of employees in the Western world is affected by work intensification.

In our own longitudinal studies, we have shown that work intensification is a stressor in addition to time pressure, leading to additional increases in burnout and to additional decreases in job satisfaction [26]. In another study [27], it was confirmed that work intensification is appraised primarily negative as an additional stressor. Only few employees perceive intensification as an ultimately positive demand. We can conclude that work intensification is a new stressor with mostly negative effects for quality of working life. With these clearly negative effects, it therefore plays no role in the context of the digital technology paradoxes.

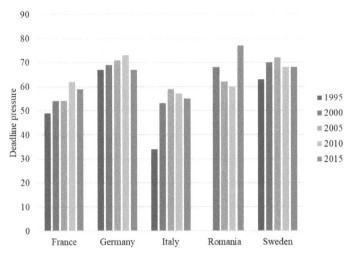

Figure 7.3 Increase of deadline pressures between 1995 and 2010 in some European countries [4].

A different picture emerges when we look at *intensified autonomy demands*. As mentioned earlier, the classic understanding of autonomy— as a resource showing positive effects when being enhanced under any circumstances—was replaced in research by a more sophisticated picture. We have showed the limits of autonomy in a larger study of elderly care employees [19]. In this study, the employees showed the highest amount of burnout both when their autonomy was perceived as low or high. Medium values of autonomy lead to the lowest values in burnout. At the same time, employees with medium levels of autonomy showed the highest values in work engagement and work motivation. Thus, autonomy demands and quality of working life indicators show a reversed U-shaped relationship. A good explanation for such a relationship is provided by Peter Warr's "vitamin model" [14], which postulates that beneficial resources at work function like vitamins. Some vitamins such as the water-soluble vitamins, e.g., vitamin C always have positive effects, which become even more positive with increasing doses, whereas other vitamins such as the liposoluble vitamin A or D have negative effects with high doses. Autonomy may work like the latter ones; thus, the aim should not be to maximize autonomy, but to find the "best" amount of autonomy for a certain job profile.

One reason for the negative effects on quality of working life of high autonomy conditions could be that high amounts of autonomy may often be found in jobs with high time pressure or little social support. Therefore, it is quite obvious that the "right" amount of autonomy is especially important in flexible working conditions. Highly mobile flexible work, where employees are able to work anytime and anywhere—and chances for social support are reduced—is a growing sector in the workforce. The vitamin model and the studies confirming this model show that the reduced quality of working life could be a price to be paid when working in highly mobile settings.

When looking at *intensified learning demands*, more complex effects on the quality of working life are also expected. There may also be challenge demands, where positive, i.e., engagement and motivation, and negative, i.e., strain reactions and reduced wellbeing, effects are observed at the same time. We studied the potential effects of intensified learning demands in a longitudinal sample of health-care workers and found positive effects of intensified learning demands over a 15-months period. An increase of learning demands was accompanied by a decrease in burnout and an increase in job satisfaction. These results were not as expected, because they confirm intensified learning demands as a resource instead of a challenge demand. At the same time, we found a positive correlation between intensified learning demands and work intensification. Combining these results, it is argued that intensified learning demands may only have predominantly positive effects until a certain amount of these demands is reached. Although when intensified learning demands occur in a work setting where also work intensification is observable, the positive motivational effects of learning may not only be lost, but could be reversed into negative effects. Then the organizational need for learning becomes a new stressor. It is once again obvious that such combinations of high intensified learning demands and high work intensification may occur in highly flexible working conditions.

7.3 Conclusion: Need-based Work Design and the Future of Work

Finally, the question, which recommendations for the evaluation of flexible work can be given arise based on the above considerations. Recommendations should be evidence based and contribute to reduce the complexity of such new demands on this topic. The following recommendations focus on work design aspects based on the personal needs of the workers.

Self-determination theory (SDT) (e.g., [28, 29]) as a useful and widely accepted theoretical framework provides an excellent base for need based job evaluation. Based on humanistic principles, SDT postulates three basic needs: the need for autonomy, competence, and relatedness. The three basic needs are culturally universal, and the satisfaction of these needs is important for every individual. Need satisfaction, both at work and in private life, is a precondition for motivation and job/life satisfaction. External motivational factors such as payments need to be accepted and internalized to play a positive role for the need satisfaction. The satisfaction of the three basic needs is a base for a good quality of working life, i.e., high job satisfaction and productivity.

There are several reasons why SDT is useful for the evaluation and design of new forms of work. For instance, in flexible working conditions, structural preconditions, such as the requirement for certain work times or the selection of a certain work place, do not exist anymore or are at least not important. However, self-organization and personal arrangements for personal working conditions are becoming more important. Employees are required to continuously decide when and where they want to work. In many circumstances, they also have flexibility in the selection of work equipment and work sequences. The needs for autonomy and competence are potentially satisfied in such working conditions. It could even be possible that new demands such as autonomy and competence demand occur as a result of an over-fulfillment of needs for autonomy and competence. A different picture emerges regarding the fulfilment of the need for relatedness. A wide range of different working times in a company and weakly coordinated working times in teams may lead to reduced social contacts or even to a feeling of loneliness at work when the commonly used formal and informal meeting opportunities do no longer exist.

Moreover, SDT may serve as an evidence-based approach to reduce the complexity at work as the appraisal of working conditions is ultimately based on the satisfaction of the three basic needs. Organizational paradoxes could also be explained by motivational factors. The needs for autonomy and relatedness can be seen as two poles of a dialectic and ultimately irresolvable tension relationship; the need satisfaction, when fulfilling one of these two needs, may reduce the satisfaction of the complemental need. However, for high job satisfaction both needs should be fulfilled as much as possible.

What would these considerations imply for practical recommendations for work design of flexible work? The actual fulfillment of the three needs could be used as a general assessment. Since flexible work allows extensive self-organization, both employees and managers are requested to design the framework requirements of their work. (e.g., job crafting). One should be careful in over-fulfilling the needs for autonomy and competence, which could lead to an increase in strain because of enhanced self-organization demands. Especially in highly flexible work environments, the framework conditions should be designed collectively. This could be, for instance, a decision to define nonwork times, e.g., no work between 10 PM and 6 AM, or the limitation of unnecessary organizational changes. Especially important in highly flexible working conditions is the design and planning of opportunities within the organization for social support and personal contacts. For instance, teams could decide to organize a regular and obligatory team meeting with clear communication rules and mutual social support measures. Leaders and managers should offer autonomy support in designing clear framework conditions and commitments and offering a maximum of autonomy within these conditions.

Even with further developments in the information and communications technology (ICT) (robotics, AI, etc.), the fulfillment of the basic needs of the workers dealing with these technologies may serve as a guideline for work design. The continuous increase in complexity, further social acceleration processes and the increase of paradoxes at work, and the resulting increase of demands could be mastered by a clear orientation on our basic needs.

References

[1] Fritz, J. and Tomaschek, N. (eds.) (2020). "Komplexe Wirkung: Digitalisierung als Triebkraft einer veränderten Arbeitswelt". Wien: Facultas Press.

[2] Perez, C. (2014). "Financial bubbles, crises and the role of government in unleashing golden ages". In: A. Pyka and H.P Burghof (Eds.) 2014; *Innovation and Finance* (pp. 11-25). London: Routledge.

[3] Internet World Stats: Usage and Population Statistics (2019). Available at www.Internetworldstats.com. [accessed July 2019].

[4] Eurofound (2016). "European Working Conditions Survey 2015". Available at https://www.eurofound.europa.eu/publications/report/2016/working-conditions/sixth-european-working-conditions-survey-overview-report

[5] Korunka, C. and Hoonakker, P. (Eds.) (2014). "The Impact of Digital Technologies in Quality of Working Life". New York: Springer.

[6] Salanova, M., Llorens, S. and Ventura, M. (2017). "Technostress: The dark side of technologies". In: C. Korunka and P. Hoonakker (Eds.) 2017; *The Impact of Digital Technologies on Quality of Working Life*. Dordrecht: Springer pp. 87-104.

[7] Rosa, H. (2005). "Beschleunigung. Die Veränderung der Zeitstrukturen in der Moderne". Suhrkamp, 161-175, Frankfurt am Main.

[8] Moore, G.E. (1965). "Cramming more components onto integrated circuits". Electronics, 38(8):114-117.

[9] Pongratz, H.J. and Voß G.G. (2003). "Arbeitskraftunternehmer: Erwerbsorientierungen in entgrenzten Arbeitsformen". Berlin.

[10] Korunka, C. and Kubicek, B. (2017). "Job Demands in a Changing World of Work".New York: Springer.

[11] Kubicek, B., Paškvan, M. and Korunka, C. (2015). "Development and validation of an instrument for assessing job demands arising from accelerated change: The intensification of job demands scale (IDS)". European Journal of Work and Organizational Psychology, 24(6): 898-913.

[12] Karasek, R.A. (1979). "Job demands, job decision latitude, and mental strain: Implications for job redesign". Administrative Science Quarterly, 24(2): 285-308.

[13] Demerouti, E., Bakker, A.B., Nachreiner, F. and Schaufeli, W.B. (2001). "The job demands-resources model of burnout". Journal of Applied Psychology, 86(3): 499-512.

[14] Warr, P. (1994). "A conceptual framework for the study of work and mental health". Work and Stress, 8: 84-97.

[15] Smith, W.K. and Lewis, M.W. (2011). "Towards a theory of paradox: A dynamic equilibrium model of organizing". Academy of Management Review, 36(2): 381-403.

[16] Hackman, J.R. and Oldham, G.R. (1976). "Motivation through the design of work. Test of a theory". Organizational Behaviour and Human Performance, 16(2): 250-259.

[17] Sewell, G. (1998). "The discipline of teams: The control of team-based industrial work through electronic and peer surveillance". Administrative Science Quarterly, 31(4).

[18] Kelliher, C. and Anderson, D. (2010). "Doing more with less? Flexible working practices and the intensification of work". Human Relations, 63(1): 83-106.

[19] Kubicek, B,, Korunka, C. and Tement, S. (2014). "Too much job control? Two studies on curvilinear relations between job control and eldercare

workers' wellbeing". International Journal of Nursing Studies, 51: 1644-1653.

[20] Leonardi, P.M., Treem, J.W. and Jackson, M.H. (2019). "The connectivity paradox: Using technology to both decrease and increase perceptions of distance in distributed work arrangements". Journal of Applied Communication Research, 38(1): 85-105.

[21] Fonner, K.L. and Roloff, M.E. (2012). "Testing the connectivity paradox: Linking teleworkers' communication media use to social presence, stress from interruptions, and organizational identification". Communication Monographs, 79(2): 205-231.

[22] Ter Hoeven, C.L., van Zoonen, W. and Fonner, K.L. (2016). "The practical paradox of technology: The influence of communication technology use on employee burnout and engagement". Communication Monographs, 83(2): 239-263.

[23] Bakker, A.B. and Demerouti, E. (2007). "The job demands-resources model. State of the art". Journal of Managerial Psychology, 22: 309-328.

[24] Eurofound and the International Labour Office (2017). "Working anytime, anywhere: The effects on the world of work". Luxembourg: Publications Office of the European Union and Geneva: The International Labour Office.

[25] Kellner, B., Korunka, C., Kubicek, B. and Wolfsberger, J. (2019). "Flexible Working Studie 2019". Wien: Deloitte.

[26] Korunka, C., Kubicek, B., Paškvan, M. and Ulferts, H. (2015). "Changes in work intensification and intensified learning: Challenge or hindrance demands?" Journal of Managerial Psychology, 30(7): 786-800.

[27] Paskvan, M., Kubicek, B., Prem, R. and Korunka, C. (2016). "Cognitive appraisal of work intensification". International Journal of Stress Management, 23: 124-146.

[28] Deci, E.L. and Ryan R.M. (2015). "International Encyclopedia of the Social & Behavioral Sciences", 2nd edition: Self-Determination Theory. Elsevier, 21, 486-491.

[29] Deci, E.L., Olafsen, A.H. and Ryan, R.M. (2017). "Self-determination theory in work organizations: The state of a science". Annual Review of Organizational Psychology and Organizational Behavior, 4: 19-43.

8

Game-changing Technologies: Impact on Job Quality, Employment, and Social Dialogue

Eleonora Peruffo[1] and Enrique Fernández-Macías[2]

[1]Eurofound, Dublin, Ireland
[2]Joint Research Centre of the European Commission, Seville, Spain

Abstract

Since the industrial revolution, economic development has been punctuated by leaps driven by the successive introduction of radical technological breakthroughs. Until recently, these breakthroughs mostly concerned the manufacturing sector, with services acting as a kind of residual category that collected all the labor displaced by technological progress. This was changed by the digital revolution, which has led to the emergence of several major technological breakthroughs with a very significant disruptive potential for both manufacturing and services in Europe. This chapter brings together qualitative research findings spanning from 2015 to 2018, which describe the potential impacts of eight different game changing technologies on work and employment in Europe. All the technologies studied rely upon and contribute to the increasing centrality of digital information in economic processes; increase flexibility and allow a mass customization of goods and services; make manufacturing and services increasingly undistinguishable; and increase significantly the efficiency and control of economic processes. In addition, these technologies often have labor-saving effects that can transform the structure of employment and tasks in Europe. This increases the relative demand of some high-skilled occupations and probably reduces labor input in physically risky or arduous tasks. However, the same technologies can have negative implications in terms of autonomy, privacy and control of workers, and conflicts related to the opacity of managerial algorithms or the ownership of the data generated in the workplace will become increasingly salient.

Keywords: Innovation, game-changing technologies, employment, technological change, working conditions, social dialogue

8.1 Introduction

Technological revolutions are historical periods of deep and fast socioeconomic change triggered by the introduction of new technologies [1][2][3]. According to [2], technological revolutions generally have two distinct phases: an "installation phase" where the new technologies are introduced and new infrastructures are built and a "deployment phase" when the new technologies are broadly adopted and their full socioeconomic transformation potential is realized. The same author considers that we are now (2019) in the deployment phase of a technological revolution brought about by digital technologies. Some of these technologies have a particular disruptive potential and can be considered as "game changing" [4][5], for instance in terms of opening up new markets or changing entire supply chains. This chapter discusses in detail eight potentially game-changing technologies (GCT) linked to the deployment phase of the digital revolution, trying to understand more in depth the implications of their adoption on a wide scale, specifically in terms of employment, working conditions, and social dialogue implications. The economic sectors under consideration are manufacturing and services.

The introduction of the internet in the late 1990s and the advent of web 2.0 in the first decade of the 21st century fostered the conditions for the application of widespread ICT-based technologies to manufacturing, creating the "networked factory" or "compu-factory" as defined in [6]. The term "Industry 4.0" refers to the widespread application of digital tools to manufacturing processes. The digitalization of manufacturing implies a deep transformation not only in terms of the technical production process, but also in terms of employment, working conditions, and work organization. In services too, digitization started in the mid-1990s and paved the way for a mutation of the economic process that is still ongoing. The use of digital tools for the coordination of service provision is facilitating entirely new forms of economic organization, as in digital labor platforms, and value generation with the increasing importance of data, as well as profound changes in work organization.

The time span over which these changes are happening is fundamental to understand these phenomena. Certainly, the speed of adoption of new technologies has accelerated in recent years; e.g., it took 25 years for telephones to reach 10% adoption, but smartphones reached a 40% penetration rate in just 10 years [7], although some infrastructural challenges might have been softened by the existence and the pre-existing experience with a telecommunication infrastructure. Moreover, this does not mean that challenges to the smooth and fast adoption of some of the technologies analyzed in this chapter are not present and, more specifically, that they are necessarily going to change the world of work in the short or medium term (or ever). For example, additive manufacturing was invented in 1983 [8] but only found actual applications in manufacturing from 2013 onwards. A reasonable time span adopted for the outlook of the selected GCTs is 10 years, i.e., up to "2027–2029".

The eight technologies selected for this study are advanced robotics, Internet of Things (IoT), electric vehicles (EV), autonomous vehicles, additive manufacturing (AM), augmented and virtual reality (AR/VR), industrial biotechnologies (IB), and blockchain.

As will be discussed throughout this chapter, the eight technologies have been selected for their disruptive game changing potential. This study looks at how these technologies can trigger changes in products or process innovations, and links these with effects on the following:

1. **employment**—potential impact on job creation and destruction, what skills could be required;
2. **working conditions**—whether these technologies are likely to affect type of contracts, working hours, and health and safety conditions;
3. **social dialogue**—whether these technologies affect social dialogue, and what is the role and response of social partners to the introduction of these technologies.

The chapter is structured as follows. First, we will contextualize our study by putting it in the context of what we think are the three main vectors of change in the digital age: automation, digitization, and coordination by platforms. Then, we will discuss the methodology followed for the study of the eight technologies analyzed. The following three sections discuss in detail our assessment of the potential implications of these technologies for the production process in manufacturing and services, employment, and industrial relations and social dialogue. Section six concludes this chapter.

This chapter brings together Eurofound's work on GCT, in particular the two reports [9] [10] on GCTs in manufacturing, part of the "Future of Manufacturing in Europe" pilot project[1], and in services, part of Eurofound's "Digital age" projects[2].

8.2 Three Vectors of Change in the Digital Age

The distinction of three main vectors of change linking digital technologies and employment, namely automation of tasks, digitization of processes, and coordination by algorithms, made in the recent publication on "Automation, digitisation, and platforms" [1] has been used as the conceptual framework for this study. These three vectors would be part of the second phase of the digital revolution as identified in [2], when infrastructure, skills, and technologies are mature for recombination and application in new fields or sectors. The GCT analyzed in this study are examples of such recombinations and applications [1] [10].

Automation, defined as the substitution of human input by machine input for some tasks, includes advanced robotics and autonomous vehicles. The novelty of these technologies in terms of automation, compared to previous applications, which were of a mechanical nature, lies in the use of algorithmic control and digital sensors, which allow machines to perform a wider range of tasks. Already some decades ago, the first digital tools (computers) allowed to automate administrative routine tasks such as paperwork and now artificial intelligence (AI)[3] may expand automation possibilities to both physical and intellectual tasks [10] characterized by nonroutine elements such as translation services or driverless cars. The ultimate frontier of automation, which would require a significant advance in existing technological possibilities, is the full automation of social tasks.

Digitization of processes refers to the use of sensors and rendering devices to translate (parts of) the physical production process into digital information and vice versa. Digitization encompasses technologies, which convert physical into virtual environments (VR), often through sensors (IoT), or enhance physical environments by superimposing information as in augmented reality (AR). The digitization of information, together with the expansion of storage tools, increases possibilities of acquiring

---

[1]https://www.eurofound.europa.eu/observatories/emcc/fome
[2]https://www.eurofound.europa.eu/topic/digital-age
[3]Machine learning and deep learning.

and analyzing information in volumes and at a speed not possible before. However, cyber-physical spaces where every object, and sometimes every person, is monitored could pose risks of erosion of privacy and autonomy in the workplace [9].

Coordination by platforms is defined as the use of digital networks to coordinate economic transactions in an algorithmic way. Airbnb and Uber are examples of platforms where the traditional role of market and business conflates into one entity: transactions are coordinated by the platform (market role) and, at the same time, these transactions are algorithmically monitored and managed by the same platform (managerial role). In relation to GCT, the coordination by platform includes technologies where coordination of processes is carried out by algorithms. The concept of smart contracts linked to blockchain applications can be used for the algorithmic coordination of work processes whereby given the fulfillment of certain conditions, a certain reward (payment/certificate and so on) can be released without human intervention. Indeed, the use of blockchain also involves automation, i.e., no human intervention is needed, and obviously digitization, i.e., all operations takes place in digital form [10].

8.3 Methodology

Two Eurofound studies on GCT, in manufacturing [9] and in services [10], were used as sources for this chapter and both applied a similar methodology. First, academic articles and gray literature on each technology were reviewed, followed by expert interviews. Finally, for the manufacturing study, workshops with practitioners and stakeholders took place, while for the services study three case studies per technology were carried out.

The point of entry for each study was a specific technology: first describing the technology uses and adoptions and then proceeding onto the analysis of the implications for employment, working conditions, and social dialogue. The time horizon for the assessment of potential implications of each technology was about 10 years (that is, up to about 2027–2029).

8.4 Impact on the Production Process

All of the technologies studied are in one way or another part of a broader process of digitalization of economic activity, referred to in the literature as the digital revolution or the digital age [1]. Following the model proposed by Carlota Pérez for characterizing technological revolutions, the technologies under consideration in this chapter would be typical exponents of the *deployment phase* of the digital revolution [2]. They rely on basic technologies and infrastructures already widely available such as microprocessors, internet, and geolocation and digital skills, which became widespread in the initial installation phase of the digital revolution. Rather than innovations as such, the eight technologies studied here are practical applications based on new combinations of those pre-existing innovations.

The fact that all the technologies studied are part of a broader process of digitalization of economic activity is very important to understand their implications for the production process and indirectly for work and employment. With only the partial exception of electric vehicles and industrial biotech, all the technologies studied here rely on essentially the same basic digital technologies: microchips, networks, sensors, big data, and algorithmic control. This means that there are very strong synergies and complementarities between the eight technologies studied, and the application of one of them could facilitate the introduction of others. For instance, introducing industrial IoT systems in a factory will enhance the standardization and control of the production process, and thus can greatly facilitate the subsequent introduction of advanced industrial robots or additive manufacturing systems. This is why these technologies are often introduced in bundles, and more generic terms such as "Industry 4.0" are used to refer to the use of these and other technologies to digitalize production processes.

Thus, the technologies studied here share some effects on the production and service provision processes, which are typical of the digital revolution [1]. Most importantly, they massively expand the availability of information on every aspect of the production and distribution processes, as well as the capability of management of making use of this information. This increases significantly the possibilities for centralized control and optimization of every aspect of the process, which not only affects the coordination of labor as we shall discuss later but also allows a more efficient use of materials and energy in production. Both in manufacturing and in services, the technologies analyzed tend to *dematerialize* parts of the production process, making the physical production of goods or provision of services increasingly secondary. For instance, additive manufacturing can collapse the entire process of physical production of a good into a single step of rendering the digital model with a three-dimensional (3D) printer; in practice, most of the value creation resides in the creation and manipulation of the digital model, which is pure information. Something similar happens with the use of virtual environments for the provision of some personal services, such as entertainment: the physical aspect of the service being completely removed, the entire transaction becomes an exchange of information.

An interesting paradox is that these common trends interact with manufacturing and services in different ways, but as a result, they tend to make them more alike. Manufacturing traditionally refers to the production of physical goods, whereas services involve different kinds of social relationships between a provider and a consumer where the product is an intangible output. The technologies studied tend to increase the centrality of information in value creation and dematerialize production also in manufacturing, thus blurring the differentiation between tangible and intangible outputs traditionally separating these sectors. With the technologies analyzed here, manufacturing is *servitized*. For instance, industrial IoT technologies allow manufacturing companies to maintain a line of communication and even control of the product after the sale, facilitating the provision of after-sales services. The physical product becomes a platform for the provision of services. The technologies analyzed can also make services more like manufacturing in some ways. For instance, advanced robotics or virtual reality environments allow for a provision of services with no direct social interaction between the producer and the consumer, so that the service becomes more like a packaged good even if it is intangible. This has been called the *productization* of services [12].

A very important way in which the technologies studied can make manufacturing and services more alike is in terms of productivity growth. It is a well-known fact of recent economic history that productivity growth tended to be much faster in manufacturing than in services, which at least partly explains the secular decline of employment in the former relative to the latter [13]. This imbalance in productivity growth is related to the different role of labor in both sectors: whereas in manufacturing labor is an input for the production of goods, in services labor is an end in itself (what is actually transacted). Thus, traditionally in many services reducing the ratio of labor input per output was either impossible or conducive to a decline in the quality of the service. Since services involve direct social relationships between producers and consumers, the possibilities for process standardization and centralized control have also traditionally been much smaller than in manufacturing. However, as mentioned earlier, some of these differences between both sectors are eroded by the technologies studied in this chapter. The availability of detailed information in real time on every aspect of the process and the efficiency of algorithmic management allow a degree of control and standardization of services, which can be very close to that of manufacturing, thus facilitating the use of tayloristic or neotayloristic forms of work organization also in services.

Interestingly, the same technologies can have the opposite impact on manufacturing, facilitating a destandardization of products that has been termed *mass customization*. In contrast with predigital Fordist mass production, which was very cost-effective but inherently rigid, these technologies open up the possibility for much more flexible production processes—thanks to algorithmic control and AI—without compromising on cost-effectiveness or process standardization. For instance, by interconnecting all objects of the production process under centralized algorithmic supervision, IoT systems increase the flexibility of the process without hampering standardization. Real time centralized control and interconnectivity allow a much faster reaction to problems, but also a relatively fast reprogramming of production in response to changes in demand or other factors.

Thus, the technologies analyzed facilitate a certain degree of convergence between both sectors, with a *servitization* of manufacturing and a *productization* of services. Some of the key traditional differences in the economic nature of both sectors are significantly eroded; not only provision processes become similarly intangible but also equally nonsimultaneous. Increasingly, the main input and source of value is information; and the massive availability of real-time data and processing capacity, together with algorithmic management, allow for similar levels of standardization and highly efficient centralized control.

8.5 Implications for Work and Employment

8.5.1 Implications for Employment

As of 2019, the technological change debate has acknowledged that technology is not only about substitution of human input but also about performance enhancement [14]. When looking at the technologies under the automation vector, **advanced robotics** and **autonomous vehicles** are the most likely to result in job substitution. While autonomous vehicles could mostly impact on the transportation sector for both goods and people, advanced robotics could impact several sectors and occupations. In particular, this study identified "four possible areas [of application]: engaging with the robots (medical and care professions), developing the robots (engineers), supervising (testing, monitoring), and new roles not existing yet" [15]. The most likely effect of the technologies analyzed is not in terms of replacing jobs, but in terms of transforming them in different ways as well as creating entirely new ones. As observed by [13] and more recently by [15], digital age technologies have led to the appearance of jobs such as "app developer, big data analysis, and software design" [16], which did not exist before.

Advanced robotics has the potential to substitute risky and hazardous jobs both in manufacturing and in services. In services, e.g., in the case of emergency workers, advanced robots could be used in situations such as natural disasters, or inspection and maintenance of dangerous sites where there is a risk of exposure to substances noxious for human beings [15]. Another use case could be police riot control, which, while protecting members of the police, could have a dehumanizing effect on the rioters and/or democracy (right to protest). On production lines and in warehouses, not only heavy items such as car parts but also delicate items such as lettuce [9] can be handled by machines whose touch and dexterity capabilities have greatly improved. One of the biggest e-commerce and logistic companies in the world, Amazon, recently announced that it will adopt a pack-wrapping machine called CartonWrap. CartonWrap is a case of automation since it can deal with 600–700 products per hour, which quadruples a worker's output in that segment. For each machine installed, only one worker is necessary to load customer orders, another to stock cardboard and glue, and a technician is required to fix jams on occasion, a requirement that Amazon would like to minimize. In terms of numbers, it takes one machine to substitute 24 workers, which for the US branch of the company means 1,300 jobs across 55 sites. The cost of a CartonWrap machine is $1.000.000 and return on investment is seen after 2 years[4].

The other technology where direct substitution of human input seems likely is **autonomous vehicles**; not only taxi drivers but also the whole transport and logistics sector, including last mile delivery, could be disrupted. After initially too optimistic projections of full automation of driving in 5–10 years, the transport industry is becoming much more cautious lately.

---

[4]Source: Reuters (2019), Amazon rolls out machines that pack orders and replace jobs. May 13, 2019. Available at https://www.reuters.com/article/us-amazon-com-automation-exclusive/exclusive-amazon-rolls-out-machines-that-pack-orders-and-replace-jobs-idUSKCN1SJ0X1

But if, or when, this were to happen, careful consideration should be given to how to provide for professional drivers losing their jobs, e.g., through reskilling programs ease their move to other economic sectors. And even if fully autonomous driving were to happen, interviews with experts carried out for this study [17] suggest that these jobs, in some cases, would not disappear but change: professional drivers' jobs not only consist in driving, but they also perform other tasks such as luggage handling, supervision and support of passengers, cleaning of vehicles, and so on. Another sector where autonomous vehicles could cause disruption is the insurance sector because it "could completely change the type of insurance policies needed, thus requiring a transformation of insurance workers' tasks and of the professional profiles of those who examine accidents who would need to have a deep understanding of algorithms and technologies used" [18]. Autonomous vehicles are likely to have an impact relatively soon on transport within private or semiprivate spaces: industrial shop floors, logistic hubs, hospitals, and university campuses are candidates for the introduction of driverless transportation. These are areas where a controlled amount of people, who could be made aware of the presence of such vehicles, circulates in fairly predictable patterns. The Rotterdam port is already using driverless vehicles to move containers, and the truck drivers have been retrained to monitor the vehicles and their operations remotely [17]. In this case the threat of unemployment has been solved through social dialogue and work organization decision.

It should not be forgotten that technologies are just tools and that working conditions crucially depend on choices at managerial and at employment relations level. From a company's perspective, automation, intended as the maximization of capital and the minimization of labor input, leads work organization toward the so-called "lights out" factory, where all processes, which can be automated will be automated. In services, this tendency seems to follow a similar pattern with what we could call a "human out" (drone deliveries, hotels without personnel, and so on) model. However, as discussed in the literature review earlier, services are intrinsically defined by the direct interaction with a user: if the user does not like the absence of human contact, some technically feasible solutions might not be adopted. It is also true that in some cases such as the introduction of automatic cashiers in supermarkets and a flight self-check-in, these practices have not provoked strong protests from consumers. These changes in the mode of service provision, which are incidentally often facilitated by the use of technologies such as the ones studied in this chapter, have at least two important implications in terms of work and employment. First, they do not involve automation as previously defined, but the replacement of (paid) workers' labor input by (unpaid) users' labor input. Second, these practices are often associated with the creation of niche markets where human interaction, with personnel devoted to improving the quality of the service by welcoming customers, giving information, and so on, is a luxury and thus available only to a small privileged minority.[5]

Under the digitization vector, applications of distributed network connectivity to the world of work can take two forms: applied to objects, e.g., to machines in industrial **IoT**, or applied to humans (**wearables**). In both cases, information gathering and real time workflow analysis and control capabilities are the main aim for adoption of this technology. For example, "virtual twins" of factories are virtual models where sensors' data is displayed to control a factory's process in near-real time. This technology is not only available in factories but also being developed and applied to aircraft maintenance, hospitals, and even cities. According to the experts interviewed, the adoption of this technology could have consequences for two types of occupations. First, monitoring occupations would require experts in AI systems and data analysis. This type of jobs would not necessarily need to be in proximity of the object being monitored, with potential consequences for local labor markets; businesses could seek these skills outside local labor market on the basis of cost considerations [6]. Second, for maintenance occupations the possibility of reducing routine inspections in favor of *ad -hoc* interventions could reduce the number of workers needed in this field.

According to our analysis, the introduction of **VR** and **AR** is also more likely to improve human performance, rather than lead to labor replacement. Beside the already commercialized use in the arts and gaming sectors, these technologies could create new jobs related to the creation of software, for instance software content for training[6] (VR) and for warehousing jobs (AR).

---

[5]Beyond the considered technologies, other forms of work such as platform work are also designed to make people work without face-to-face contact and limiting people's movement, working from home, getting food delivered at home by a drone, and managing people through algorithms.

[6]Early applications of VR training have been utilized by the defence forces and by emergency and rescue workers. VR allows training to take place in a "safe space" and the exercise can be repeated to deal with errors, while guidance can also be incorporated in the training. Pilot projects have shown that these trainings bear the same result as live ones in terms of learning, although a real-world experience should complement the virtual one for the type of multisensorial stimuli, which even a very sophisticated software cannot yet mimic.

Further, AR could have a substitution effect in the tourism sector where information about surroundings such as tourist sites can be performed by AR instead of a tour guide.

Preliminary data on **additive manufacturing** employment effects do not point to job losses. And again, the preliminary analysis of cases of adoption of this technology hints at a change in work processes where tasks could be transformed or new tasks could be created. Additive manufacturing is mostly used for fast prototyping and in sectors where innovative shapes are part of the creative process (arts and fashion, jewellery). In manufacturing, possible new roles around additive manufacturing could be those of machinery maintenance and cleaning of production spaces.

Media hype on **blockchain** has probably just passed its peak at the time of writing. Despite all the promises, at the moment it is mostly used in finance for cryptocurrencies and the odd case of smart contracts, such as the blockchain implementation by the singer Imogen Heap for selling songs directly to consumers (the song is sold to the user when the blockchain records the money transfer) [19]. Possible fields of application vary from education certificates to value chain control, and potential employment impacts would mostly affect intermediaries' occupations such as administrative staff or notaries where employees are in charge of validating certifications or contracts. The European Commission lists more than 600 pilot projects on its blockchain observatory, and many other initiatives are taking place in private and public settings. Certainly, these pilot projects created jobs in software and engineering, and for legal teams in charge of compliance; if successful, financial institutions' back offices could reduce their workforce following the implementation of blockchain workflows[7].

8.5.2 Implications for Working Conditions

Work organization plays a mediating role with respect to the impact of the technologies analyzed on the different job quality dimensions, namely, intrinsic quality of work, employment quality, workplace risks, and working time and work life balance. In this study, we used Eurofound's job quality framework [20] to analyze these impacts.

On the one hand, these types of trainings offer a cost reduction opportunity. For example, there is no need to burn a building every time; on the other hand, the headset equipment and the cost of a bespoke software could be high.

[7]https://www.santander.com/csgs/Satellite/CFWCSancomQP01/en_GB/Corporate/Press-room/2019/09/12/Santander-launches-the-first-end-to-end-blockchain-bond.html

Intrinsic quality of work includes skills, autonomy, and social support. Skills emerged as one of the most important job quality dimensions in the analysis of the technologies. But what are the emerging skills needs arising from the studied technologies?

According to the ten technology reports, a high level of ICT skills will be required for jobs where workers will have to deal with advanced robotics, IoTs, and additive manufacturing, in particular data analytics, software engineering and mechatronics as well as technical skills for digital equipment maintenance. But in terms of future skill needs, our findings corroborate the literature [21] [22] not only for what concerns an increasing need of ICT skills but also for social skills such as the ability to communicate within an interdisciplinary team. For example, to run complex industry 4.0 factories, a new mix of profiles from traditional engineering and manufacturing skills to new digital-related skills, such as big data management, is required, thus implying increased needs of communication and teamwork across different areas.

According to the European Jobs Monitor 2013 [20], **employment quality** is formed by two subdimensions: "the degree of contractual stability of the worker" and "prospects that the job provides for the further development of the employee." The subdimension of contractual stability is linked to that by skills; the higher the specialization in ICT-related skills the higher the chance for the worker to maintain a job, be it in manufacturing or in services. Nevertheless, this pattern might be disrupted by industry 4.0, i.e., advanced robotics, IoT, and additive manufacturing, and the increase in remote working possibilities offered by the digitization of work such as, e.g., telemedicine. This might enable shorter tasks and on-demand services leading to a gig economy with no stability nor prospects rather than continuous provision, at the same time opening labor markets to global competition for jobs, which were previously linked to local presence. It could be argued that, not all, but some of these tasks might still be quite specialized so a niche for intellectual task for highly skilled workers could see an increase in self-employment. In terms of labor market disruption, there are policy decisions such as certification and guild membership, which can protect salaries from global competition and also ensure customer protection by setting standards.

In terms of **workplace risks** both in manufacturing and in services, a decline of physically demanding jobs resulting in physical strain is a result of advanced robotics. Heavy items or risky situations can be handled by purpose-built robots such as machines moving and moulding car parts or robots carrying out inspections in emergency. Musculoskeletal disorders could be alleviated by the use of exoskeletons in manufacturing, while in transport, where long hours in the same position can cause health issues to drivers, could be diminished by highly automated vehicles.

Hazardous workspaces can become safer if monitored through sensors, e.g., by sensors alerting of poisonous substances in the air or by monitoring workers' location or heart rate [18]. These advantages should, however, be treated with caution since monitoring and control are divided by a thin line; workers could be controlled beyond stated purposes and information gathered about them without explicit consent [23].

Materials used in additive manufacturing could have an unknown repercussion on workers' health: new powders and pastes have not been in use long enough to fully know all the associated hazards. For these reasons, additive manufacturing spaces should be built [24] in well-aerated areas. At the same time, additive manufacturing could lead to cleaner spaces, due to the fact that materials' residues are limited and to an increase of automated processes, since product design and prototyping can be adjusted faster.

With respect to **working time and work–life balance**, data collection on production processes and on services' interactions can lead to a rearranging of traditional work's schedules. In services, customer demands, or customers visits to a certain shop, can be better predicted thus allowing companies to concentrate workers shifts into the busiest slots. In manufacturing, while *ad hoc* interventions enabled by sensors' data monitoring might facilitate a reduction of working hours, these *ad hoc* interventions come with a potential downside, which is 24/7 availability possibly disrupting workers' work–life balance and loss of autonomy. Teleworking could also be enabled by automation and digitization with advantages such as reduced commuting time and organizing one's own working schedule; but in some cases, this could lead to disadvantages such as stress derived by on-demand availability [25].

8.6 Impact on Industrial Relations and Social Dialogue

Social dialogue has the potential to be a central instrument to manage technological change in the digital age. In the interviews carried out for this study, the most frequently mentioned potential impact of the studied technologies on industrial relations at the workplace is, perhaps surprisingly, related to automation rather than digitalization as such. Automation technologies including advanced robotics and autonomous vehicles tend to replace labor input for low and mid-skilled blue-collar workers, the traditional constituency of unions, and thus could undermine the balance of power at the workplace. Some of the interviews pointed to IoT as a highly transferrable technology, which can be used to speed up automation of processes and monitoring of workers. The capability of collecting workers' data through sensors, even personal data, raises issues about the ownership of that data, which should be discussed and agreed by social partners or otherwise regulated by the government. Efforts should be made to avoid a scenario where detailed information about workers—not only about performance, but potentially including even personal biodata—is the exclusive property of company, which would give an unfair and excessive advantage to managers with respect to workers at the workplace level.

The involvement of HR departments in technological change is also directed toward policies which fast-track automation, but this should not leave aside the way in which workers engage with technology. Social dialogue should aim at policies, which also take into account the impact of technological change on different age cohorts: typically, older workers need reskilling when new technologies are introduced. So far, there is limited evidence of impacts of blockchain on social dialogue but the potential of algorithmic coordination is likely to favor a work environment where freelance and distributed work are increasingly prevalent, a situation which would also lead to a weakening of organized labor in the workplace [26].

8.7 Conclusion

How disruptive will be the eight technologies studied in this chapter, and when will they have a significant effect on work and employment? As of 2020, it is challenging to give a decisive answer or a precise timeframe. The terminology might need to be revised from disruption to progressive implementation in most cases. From the findings of this study, it certainly seems that some changes be expected in the services sector in the short- (robotics in emergency and rescue), medium- (wearables and VR/AR), and long term (autonomous vehicles). In these cases, the most obvious effect is the improvement in workers' health and safety or the reduction of physical effort. In manufacturing, advanced robotics is being increasingly applied beyond the traditional sectors of oil, aeronautics, and automotive into sectors where dexterity and material of different textures need to be processed such as textiles or shoes production. In services, the 2020 COVID-19 pandemic has prompted acceleration in health-care robotics initiatives due to the necessity of reducing human-to-human contact; but despite some successful use cases like the use of disinfecting robots in hospitals, the media hype of recent year has been deflated, wide adoption is still a few years ahead. The vast applicability of IoT in both manufacturing and services is already resulting in a digital transformation not only of companies but of value chains too. Additive manufacturing is still marginal as an actual manufacturing technology, but it is gaining importance as a tool for fast prototyping and so is VR/AR for collaborative design and in logistics. In all the cases studied, adoption is likely to be uneven and dependent on specific factors at company or sector level. Companies are interested in and want to adopt new technologies but the recombination phase is ongoing, early adopters are experimenting and the majority is waiting for results, and studying how their activity can adapt or benefit from the adoption of a certain technology (or set of) technologies. Moreover, investments costs are measured against still uncertain benefits. The 2020 pandemic might change completely scenarios which looked set until a few months earlier, companies could decide to expand their workforce due to high availability of labor and postpone technology investment when goals become clearer. Or, due to health and safety measures which at the time of writing require physical distancing, companies might decide to reduce their workforce and automate their processes and services as soon as possible with the aim of reducing risks of employees falling ill, and avoid organizational and perhaps legal consequences.

In terms of employment, the most frequently cited potential effect of these technologies is job losses for low- and mid-skilled workers thus prompting for retraining and upskilling policies in an effort to mitigate the social impact of job losses due to technological change. But this study also pointed toward new job opportunities for professionals specialized in data analytics, software, engineering, and maintenance. There will certainly be additional job opportunities associated with these technologies that we can simply not foresee yet or changes in job tasks that may require workers to develop very different skills. The natural experiment of mass telework during spring 2020, as observed by many, might have provoked a monumental shift in the way work is organized and may have accelerated the digitalization of certain tasks, although to what extent this is going to endure is hard to assess at present. These different trends in labor demand will imply changes in the demand for skills also, most likely with a rising demand for advanced digital skills as well as interdisciplinary and communication skills.

Both positive and negative aspects have been observed for working conditions. The diffusion of automation and digitization technologies could favor teleworking but also a fragmentation of job tasks and 24/7 demands, which could infringe on work–life balance. Another aspect already observed is the shift of labor from workers to customers, which can also lead to job losses without actually automating any work process as such. The way in which work is organized also has an impact on how technology is implemented, and on how working conditions and employment is affected in the end. Choices at organizational level should be discussed among social partners to find mutually satisfying solutions, especially in terms of monitoring and data usage.

8.8 Acknowledgments

The authors would like to thank Eurofound colleagues Irene Mandl, Donald Storrie, and Ricardo-Rodríguez Contreras for their comments and feedback. Thank you also to Martijn Poel and the Technopolis team, and to Salil Gunashekar and the RAND team.

References

[1] Fernández-Macías, E. (2018). "Automation, Digitisation and Platforms: Implications for Work and Employment". Working paper, Eurofound, Dublin. [Online]. Available at https://www.eurofound.europa.eu/sites/default/files/wpef17035.pdf Accessed date: 8/01/2020

[2] Perez, C. (2002). "Technological Revolutions and Financial Capital: The Dynamics of Bubbles and Golden Ages". Cheltenham: Edward Elgar.

[3] Zuboff, S. (2010). "Creating value in the age of distributed capitalism", McKinsey Quarterly, 4, 44–55. [Online]. Available at https://www.mckinsey.com/business-functions/strategy-and-corporate-finance/our-insights/creating-value-in-the-age-of-distributed-capitalism# Accessed date: 8/01/2020

[4] Bower, J.L. and Christensen, C.M. (1995). "Disruptive Technologies: Catching the Wave", Harvard Business Review [Online]. Available at https://hbr.org/1995/01/disruptive-technologies-catching-the-wave?referral=03759&cm_vc=rr_item_page.bottom Accessed date: 8/01/2020

[5] OECD and Eurostat (2018). "Oslo Manual 2018". OECD.

[6] Baldwin, R. (2017). "The Great Convergence: Information Technology and the New Globalization". Cambridge, Massachusetts: Belknap Press of Harvard University Press.

[7] McGrath, R.G. (2013). 'The Pace of Technology Adoption is Speeding Up". Harvard Business Review [Online]. Available at https://hbr.org/20 13/11/the-pace-of-technology-adoption-is-speeding-up Accessed date: 7/12/2020

[8] Peruffo, E., Rodriguez Contreras, R., Molinuevo, D. and Schmidlechner, L. (2017). "Digitisation of Processes". Working paper, Eurofound, Dublin. [Online]. Available at https://www.eurofound.europa.eu/site s/default/files/wpef17038.pdf Accessed date: 7/12/2020

[9] Eurofound (2018). "Game changing technologies: Exploring the impact on production processes and work", Publications Office of the European Union, Luxembourg. [Online]. Available at https://www.eurofound.eu ropa.eu/sites/default/files/ef_publication/field_ef_document/fomeef18 001en.pdf Accessed date: 7/12/2020

[10] Eurofound (2020). "Game-changing technologies: Transforming production and employment in Europe", Publications Office of the European Union, Luxembourg. [Online]. Available at https://www. eurofound.europa.eu/sites Game-changing technologies: Transforming production and employment in Europe/default/files/ef_publication/ field_ef_document/fomeef18001en.pdf Accessed date: 7/12/2020

[11] Fernández-Macías, E. (2019). "Three vectors transforming work in the digital revolution", Eurofound. [Online]. Available at https://www.euro found.europa.eu/publications/blog/three-vectors-transforming-work-in -the-digital-revolution Accessed date: 8/01/2020

[12] Valminen, K. and Toivonen, M. (2009). "Productisation of Services: What , Why and Howă?" In 14th International Conference of RESER Conference, Budapest, 2-5, Sept. 24–26.

[13] Baumol, W.J. (1967). "Macroeconomics of Unbalanced Growth: The Anatomy of Urban Crisis", The American Economic Review, 57(**??**), 415–426. [Online]. Available at https://www.jstor.org/stable/1812111?o rigin=JSTOR-pdf&seq=1 Accessed date: 7/12/2020

[14] Autor, D.H. (2015). "Why Are There Still So Many Jobs? The History and Future of Workplace Automation". Journal of Economic Perspectives, 29(**??**), 3–30. [Online]. Available at https://pubs.aeaweb. org/doi/pdfplus/10.1257/jep.29.3.3 Accessed date: 8/01/2020

[15] Eurofound (2019). "Advanced Robotics: Implications of Game-changing Technologies in the Services Sector in Europe". Working paper, Eurofound, Dublin. [Online]. Available at https://www.eurofo und.europa.eu/sites/default/files/wpfomeef18003.pdf Accessed date: 7/12/2020

[16] Berger, T., Chen, C. and Frey, C.B. (2018). "Drivers of Disruption? Estimating the Uber effect", European Economic Review, 110, 197–210. [Online]. Available at https://www.sciencedirect.com/science/article/pii/S0014292118300849 Accessed date: 12/09/2020

[17] Eurofound (2019). "Autonomous Transport Devices: Implications of Game-changing Technologies in the Services Sector in Europe". Working paper, Eurofound, Dublin [Online]. Available at https://www.eurofound.europa.eu/sites/default/files/wpef19002.pdf Accessed date: 7/12/2020

[18] Eurofound (2019). "The Future of Manufacturing in Europe", Publication Office of the European Union, Luxembourg. [Online]. Available at https://www.eurofound.europa.eu/sites/default/files/ef_publication/field_ef_document/fomeef18001en.pdf Accessed date: 7/12/2020

[19] Tapscott, A., Tapscott, D. and Cummings, J. (2016). "Blockchain Revolution". New York: Penguin.

[20] Eurofound (2013). "Employment Polarisation and Job Quality in the Crisis: European Jobs Monitor 2013". Publication Office of the European Union, Luxembourg. [Online]. Available at https://www.eurofound.europa.eu/sites/default/files/ef_publication/field_ef_document/ef1304en.pdf Accessed date: 12/09/2020

[21] Frey, C.B. and Osborne, M.A. (2013). "The Future of Employment: How Susceptible Are Jobs To Computerisation?" Working paper, Oford University. [Online]. Available at https://www.oxfordmartin.ox.ac.uk/downloads/academic/The_Future_of_Employment.pdf Accessed date: 12/09/2020

[22] Cedefop (2018). "Insights into Skill Shortages and Skill Mismatch: Learning from Cedefop's European Skills and Jobs Survey", Luxembourg: Publications Office of the European Union. [Online]. Available at https://www.cedefop.europa.eu/files/3075_en.pdf Accessed date: 12/09/2020

[23] Zuboff, S. (2019). "The Age of Surveillance Capitalism: The Fight for a Human Future at the New Frontier of Power". New York: PublicAffairs.

[24] Eurofound (2018). "Additive Manufacturing: A Layered Revolution. Impact of Game-changing Technologies in European Manufacturing". Working paper, Dublin. [Online]. Available at https://www.eurofound.europa.eu/sites/default/files/wpfomeef18002.pdf Accessed date: 12/09/2020

[25] Eurofound and International Labour Office (2017). "Working Anytime, Anywhere: The Effects on the World of Work". Publications Office of the European Union, Luxembourg, and the International Labour Office, Geneva. [Online]. Available at https://www.eurofound.europa.eu/sites/default/files/ef_publication/field_ef_document/ef1658en.pdf Accessed date: 09/01/2020

[26] Johnston, H. and Land-Kazlauskas, C. (2019). "Organizing On-Demand: Representation, Voice, and Collective Bargaining in the Gig Economy", ILO Working Paper, Conditions of Work and Employment Series, No. 94. [Online]. Available at https://ideas.repec.org/p/ilo/ilow ps/994981993502676.html Accessed date: 09/01/2020

9

The Diversity of Platform Work— Variations in Employment and Working Conditions

Irene Mandl[1] **and Cristiano Codagnone**[2]

[1]European Foundation for the Improvement of Living and Working Conditions (Eurofound), Dublin, Ireland
[2]University of Milan, Milan, Italy

Abstract

Platform work emerged as an employment form and business model in Europe about a decade ago. While it is still small in scale, it is dynamically developing. This also refers to an increasing heterogeneity within platform work, which results in different effects on employment and working conditions of platform workers. This chapter suggests a classification of platform work using a combination of five criteria: scale of tasks, skills level required to fulfill them, format of service provision, selector of task assignment, and form of matching. Applying these criteria identifies 10 distinctive types of platform work which had some critical mass as of 2017. Based on this, the chapter discusses the employment and working conditions of platform workers affiliated to 5 of these 10 types. It stresses that there is no type of platform work which exclusively poses advantages or disadvantages to the workers. Indeed, their opportunities and risks vary quite substantially. That said, platform work which is related to small-scale, low-skilled tasks (algorithmically) assigned to the worker by the platform which—beyond matching—also determines work organization tends to raise more challenges for workers and the labor market.

The chapter concludes with policy pointers, flagging the need for a more differentiated policy approach which better considers the heterogeneity in platform work.

Keywords: Platform work, platform economics, digitalization, working conditions, employment status, job quality, labor market

9.1 Introduction

In this introductory section, we develop a few arguments on how online labor platforms are positioned with respect to the economic work on two-sided markets, present a working definition and typological distinction, and extract from the economic literature the *ex ante* hypotheses about their potential impacts. These general considerations and hypotheses are then triangulated with more nuanced empirical evidence about types of platform work (section 2) and their impacts on work and employment (section 3). We conclude with a few policy pointers.

Online platforms intermediating labor-intensive services, i.e., Upwork, Uber, Amazon Mechanical Turk, Task Rabbit, just to name a few, have been surrounded by controversies and rhetorical disputes. They are alternatively presented as a source of opportunities for people and of market efficiency or instead of prevarication (for a full review, see Refs. [1, 2].

As it will become clear once we discuss the literature on two-sided markets, these platforms are hybrids between market and hierarchy rather than pure forms of a two-sided market. This aspect is relevant, for instance, when considering the legal dispute about the status of individuals performing tasks through labor platforms as simple contractors or de facto employees. If online platforms exert a large degree of control as in a hierarchy, then it is more difficult to claim that individuals performing the labor tasks are self-employed contractors [3, 4, 5]. Whether or not a particular activity qualifies as a two-sided market has relevant implications also with respect to competition policy [6, 7, 8, 9].

Economists have analyzed platforms such as Uber, Airbnb, Upwork, and TaskRabbit as two-sided markets, adding a labor market perspective to the more generally considered economic mechanism angle by discussing the issue of control over the service provider [3, 4, 5, 10, 11, 12, 13, 14, 15, 16].

On two-sided market-like platforms, two groups of users transact or interact in ways that create positive network externalities, as it has been analyzed since 2002 by a growing body of economic literature

[6, 9, 17, 18, 19, 20, 21, 22, 23, 24, 25, 26, 27, 28, 29], but the conditions for two-sidedness (or multisidedness) still remain an empirical matter to be ascertained case by case [30, 31].

In the original treatment of two-sided market-like platforms, two-way network effects were indirectly considered a necessary condition [18, 27, 28]. Subsequently, other authors relaxed this condition and consider sufficient that one side benefits from network effects for the qualification as two-side market-like [8, 20, 32]; and more recently [30]. According to Ref. [20], the conditions for a two-sided market-like platform are that (a) there are distinct groups of customers, (b) a member of one group benefits from having demand coordinated with one or more members of another group, and (c) an intermediary can facilitate that coordination more efficiently than bilateral relationships between the members of the group. Later, Evans and Schmalensee [8] added that for condition (b) above, it is sufficient that one side of customers is attracted with the increasing size of the other. In their second contribution, Rochet and Tirole [29] proposed the following definition, where the key and only characterizing element is apparently that the price structure is nonneutral: "*A market is two-sided (a two-sided platform exists) if the platform can affect the volume of transactions by charging more to one side of the market and reducing the price paid by the other side by an equal amount; in other words, the price structure matters, and platforms must design it so as to bring both sides on board*" ([29], pp. 664–665).

On the other hand, Hagiu and Wright [3, 4, 5, 14] look at two- or multi sidedness as a matter of firms' strategic choices. Building on the theory of the firm, they frame these choices as a trade-off between being a two-sided or multisided platform or a vertical integrated firm [4], or between controlling versus enabling [5]. They contrast two-sided platforms to vertical integrated (VI) firms or resellers alongside the dimension of the amount of control exerted by the different players. In this perspective, the key features of two-sidedness and multisidedness are that (i) they enable direct interactions between two or more distinct sides and that (ii) each side is affiliated with the market/platform. Direct interaction entails that the parties maintain control over key terms of the interaction, i.e., pricing, bundling, delivery, marketing, quality of the goods or service offered, and terms and conditions, as opposed to a situation where the intermediary takes control over such terms.

In view of the above discussion, we can clear the field from the noise of various names used for the phenomenon considered here, i.e., "sharing economy," the "collaborative economy," "crowd-working," "crowd-sourcing," the "gig economy," and the "on-demand economy," and state that online labor platforms are two-sided markets as long as no part exerts excessive control. When control in practice is salient then they move toward a hybrid form between market and hierarchy and, justifiably, give rise to the legal disputes on workers' employment status that sprung up most everywhere in the last 3 to 4 years.

Based on several contributions [33, 34, 35, 36], we adopt as a starting point the following definition of online labor platforms:

> *(1) work as digital marketplaces for non-standard and contingent work; (2) where services of various nature are produced using preponderantly the labor factor (as opposed to selling goods or renting property or a car); (3) where labor, i.e., the produced services, is exchanged for money; (4) where the matching is digitally mediated and administered although performance and delivery of labor can be electronically transmitted or physical; (5) where the allocation of labor and money is determined by a collection of buyers and sellers operating within a price system* ([1], p. 74 and pp. 76-83).

Within this definition, one can make a first high-level distinction using two key dimensions: (a) whether tasks are entirely traded and delivered online, or they are traded, monitored, and paid online, but the delivery is physical and collocated and (b) what type of tasks are traded and what skills are required to deliver them, i.e., low skills mostly routine or manual vs high skills and mostly cognitive and interactive. On the basis of the first dimension, we can distinguish online labor markets that are potentially global from mobile labor markets (MLMs) that are by definition localized. The second dimension distinguishes between intermediation of tasks embedding a relatively lower or relatively higher skills set. More fine-grained distinctions are presented in the next section. Three main general policy relevant aspects can be drawn from this definition and the related distinction.

First, online labor markets are global and difficult to regulate as the platform, the client and the worker may reside in different countries, whereas mobile labor markets are localized at least as regards the client and the worker and an easier target for regulation. Second, considering skills, content of labor tasks sold and bought in online and mobile labor market enables to tap into the debate on the routine biased technical change (RBTC) hypothesis [37, 38, 39, 40, 41]. This also has a clear regulatory relevance in that very high skilled contractors maybe at ease with working as freelancers, whereas platform workers involved in low level routine cognitive tasks or in manual work maybe in need for more social and employment protection. Third, several of the most well known labor platforms exert a level of control to shed serious doubts on whether they qualify as two-sided market or reseller firms.

Moving away from definition issues, an additional aspect to introduce is the expected impact that these labor platforms can have on the labor market. Based on the hypotheses from the economic literature, the following effects are expected: (a) increased labor market participation for specific groups (women, students, older workers, migrants); (b) flattening of labor markets and increased meritocracy; (c) distributional effects, i.e., prevalence of superstar vs long tail effects; and (d) net aggregate effects, i.e., the increased efficiency of labor market matching and the increased production efficiency due to lower coordination costs. The evidence from experimental and observational economic studies is inconclusive on most of these effects. There is some evidence to support the growing female participation [42, 43, 44]. Some studies suggest that online platforms do not make labor markets as flat as expected because geographical, cultural, and language differences still matter [45, 46, 47, 48, 49, 50, 51]. Studies focusing on online labor markets find concentration of work assignments (if not full blown "superstar" effects) and no case of long tail effects [14, 34, 52, 53]. The evidence available does not warrant any conclusion about net aggregate effects ([54], pp. 23–25).

9.2 Types of Platform Work in Europe

While platform work is still small in scale in Europe, it has been dynamically developing during the last decade. This not only refers to its scale, but also its scope. The types of tasks mediated through online platforms or apps, as well as the business models of platforms, are getting more and more diverse. The reason for such an increasing heterogeneity seems to lie, on the one hand, in a growing acceptance of clients to use platforms to find/access service providers. On the other hand, due to the inherent network effect of the platform economy, platforms who are not among the "top players," and notably new market entrants, need to offer something different from existing platforms to attract and retain clients and workers. Both results in platform work being used for a larger variety of tasks, but also in changing business models (e.g., as regards to ancillary services offered by the platforms to clients or workers).

Against this background, the empirical evidence commented in this section brings important nuances to the initial definition and typological distinction made in the introduction. A wide variety of different elements to classify platform work more granularly can be considered [55]. Eurofound [56] proposed to focus on five of them, which are deemed to be the most important ones, as well as benefitting from some data available to indicate their relevance. The suggested classification elements are the following:

- the skills level required to perform the task (low, medium, or high);
- the format of service provision (on-location or online);
- the scale of the tasks (microtasks vs larger projects);
- the selector (decision made by the platform, client, or worker); and
- the form of matching (an offer or a contest).

Applying these categorization elements results in 120 potential types of platform work. When feeding these combinations with available data, it emerged that as of 2017, there were 10 types of platform work in Europe that had some critical mass in terms of the number of platforms and affiliated active workers (Table 9.1).

Such a classification is important as the various types of platform work show considerable differences in the employment and working conditions of the affiliated workers, as will be shown in the next section. An important influence factor is the position where the type of platform work stands in the hierarchy-to-market dichotomy (Fig. 9.1): As mentioned in the previous section, platforms are often described as a hybrid organization form, situated between "market," i.e., with autonomous market players defining their own terms and conditions and discretion over business decisions, and "hierarchy," i.e., with strong elements of subordination and dependency relationships. To give just a few examples of the above types, from a labor market perspective, the on-location platform-determined routine work and the online moderately skilled click-work tend to resemble more hierarchical organizations as the platforms often provide instructions and exert algorithmic control [2]. On-location client-determined and worker-initiated moderately skilled work, or online contestant specialist work in contrast, are more at the market end of the spectrum, acting as a matching tool with limited interference in the actual service provision.

Another important influence factor for working conditions is who decides on task assignments and the terms and conditions of the service provision. For example, in the on-location worker initiated moderately skilled work the worker decides which offers received from potential clients to accept and has high discretion over prices and work organization. In the on-location platform-determined routine work, in contrast, the platform generally decides

Table 9.1 Most common types of platform work in the EU, 2017 ([56] based on the JRC database; [57, 58]).

Labels	Service Classifications			Platform classifications		Shares of platforms in total number of platforms	Shares of workers in total number of workers	Examples
	Skills level	Format of service provision	Scale of tasks	Selector	Form of matching			
On-location client-determined routine work	Low	On-location	Larger	Client	Offer	13.7%	1.3%	GoMore (transport)
On-location platform-determined routine work	Low	On-location	Larger	Platform	Offer	31.5%	31.2%	Uber (transport)
On-location client-determined moderately skilled work	Low to medium	On-location	Larger	Client	Offer	11.3%	10.9%	Oferia (household services)
On-location worker-initiated moderately skilled work	Low to medium	On-location	Larger	Worker	Offer	4.2%	5.5%	ListMinut (household services)
Online moderately skilled click-work	Low to medium	Online	Micro	Platform	Offer	0.6%	5.3%	CrowdFlower (professional services)

(Continued)

Table 9.1 Continued

Labels	Service Classifications			Platform classifications		Shares of platforms in total number of platforms	Shares of workers in total number of workers	Examples
	Skills level	Format of service provision	Scale of tasks	Selector	Form of matching			
On-location client-determined higher-skilled work	Medium	On-location	Larger	Client	Offer	2.4%	3.3%	appJobber (professional services)
On-location platform-determined higher-skilled work	Medium	On-location	Larger	Platform	Offer	1.2%	4.2%	Be My Eyes (professional services)
Online platform-determined higher-skilled work	Medium	Online	Larger	Platform	Offer	0.6%	1.9%	Clickworker (professional services)
Online client-determined specialist work	Medium to high	Online	Larger	Client	Offer	5.4%	30.3%	Freelancer (professional services)
Online contestant specialist work	High	Online	Larger	Client	Contest	5.4%	4.6%	99designs (professional services)

Figure 9.1 Hierarchy-to-market dichotomy by analyzed type of platform work [2].

on task assignment—in most cases through algorithmic matching—and imposes work organization and working conditions on the workers.

9.3 The Impact of Platform Work—Diversity Across Types

In public and policy discussions on platform work, mainly its negative effects are illustrated, such as low income, unpaid working time, unpredictability of task assignments, insecure work environments, or unfair ratings. While these are certainly true for some types of platform work, they do not manifest in all types. There are also opportunities emerging from platform work—both for the individual worker and the overall labor market. A more nuanced approach toward platform work is needed to present a more realistic picture about the phenomenon.[1]

9.3.1 The Macro Perspective: Platform Work and the Labor Market

One of the main advantages widely discussed for platform work—see the first hypothesis also derived from the extensive literature in section 1—are its capacities to facilitate labor market participation due to its low entry barriers in terms of required proofs of qualifications and related administrative

---

[1]The following is based on Eurofound [2, 56]. For more details and further elaboration, please refer to the reports, as well as to further information on the various topics mentioned as gathered in Eurofound's web repository on the platform economy, https://www.eurofound.europa.eu/data/platform-economy.

procedures. This holds true for both broad categories of platform work—the one which resembles more a hierarchy and the one which is more market based. Going into more detail, however, the analysis of Eurofound [2, 56] shows that this holds particularly true for those types of platform work that are linked to low skills requirements (Fig. 9.2). Furthermore, those tasks that are delivered online, irrespective of the level of skills requirement, can be conducted independent of a specific location, hence supporting labor market participation of those in remote or disadvantaged geographic areas. For both groups, this also results in an opportunity to comparatively easily and quickly earn (additional) income. Accordingly, one could confirm the hypotheses on facilitated labor market access and increased productivity and efficiency raised in section 1.

On the negative side, however, is that this type of work poses some risk to crowd out traditional employment with better working conditions, and leaves workers who would prefer more traditional employment "stuck" in platform work, hence contributing to labor market segmentation. Also, as the available data hint toward platform workers tending to be highly qualified but engaging (continuously and exclusively) in low-skilled tasks, it bears some risk of deskilling. This not only harms the individual worker, but can also be of a wider disadvantage as it leaves available potentials untapped or underutilized for the benefit of the economy. This risk is more prominent for low-skilled tasks mediated through platforms which exercise a high level of control, that is, those that act more as hierarchical organization than just matching supply and demand for labor.

For some of the types of platform work related to higher skilled tasks, and those that act more on a market-based model, it can be observed that workers use this employment form strategically to either enhance their already existing self-employed activity that is, use platforms to find additional clients to bridge idle times, or to try out self-employment with a comparatively low level of risks involved. From a macroeconomic perspective, such can positively contribute to the entrepreneurial spirit in the economy, which, in turn, can benefit the competitiveness and level of innovation in the country or region.

The types of platform work that are characterized by a rather hierarchy-like business model, show some likelihood of misclassification of the employment status of the worker if they are considered self-employed or freelancer (and hence are subject to entrepreneurial risk) while they do not enjoy the autonomy and flexibility of being one's own boss. This leaves the individual worker in a worse-off situation as regards rights and entitlements

Figure 9.2 Potential labor market effects of different type of platform work [2].

for working conditions, social protection, and representation than if they were properly classified.

9.3.2 The Micro Perspective: Platform Work and Working Conditions

Algorithmic management is among the key characteristics of platform work. This can be to the benefit of the worker if the algorithm contributes to increased efficiency and objectivity in selecting/assigning tasks, particularly e.g., in online click-work, which is dominated by workers who tend to belong to the disadvantaged groups in the labor market. Accordingly, the hypothesis on increased efficiency derived in section 1 could be confirmed while the one related to unfair distributional effects could be dismissed.

However, if the algorithm is unfavorably programmed, it impedes the workers' flexibility and autonomy by exerting a high-level control or influencing their access to specific types of tasks without being transparent about that to the worker, which negatively affects working conditions. There is a higher likelihood for such to happen in hierarchy-like platform business models and low-skilled tasks compared to other platform work types. For these, the hypothesis on unfair distributional effects could be confirmed (Fig. 9.3).

Platform work is generally widely praised for its inherent working time flexibility. This is confirmed for those types of platform work where higher skills are required and where the platform applies a market-like business model, hence giving the worker more discretion to agree with the client on work organization, including working time. In hierarchy-like business models, there is some tendency of assigning tasks on short notice, requiring workers to be available "anytime" to respond quickly if they want to avoid losing work opportunities. Online contests are at least partly characterized by higher work intensity as deadlines tend to be shorter than for comparable tasks in the traditional economy. Hence, while the effects for the workers differ depending on the type of platform work, the findings generally confirm the hypothesis on increased efficiency in labor market matching and service provision.

Similarly, the issue of earnings deserves a more differentiated discussion. While low-skilled tasks are linked to low earnings—due to their small scale and low requirements—payments tend to be decent in terms of comparable to rates in the traditional economy if services are delivered on-location. In contrast, online services show some higher probability of being paid less than in the traditional economy due to the "faceless" nature and higher (global) competition. Furthermore, online tasks are often characterized by a rather high level of unpaid time needed by the worker to search and bid for tasks, which might not be assigned to them. As regards the predictability of earnings, this tends to be rather good in platform-determined work while work opportunities, and hence income, is quite unpredictable in online click-work and online contests, again mainly due to the potential global competition. As regards the differentiation along the hierarchy-to-market-dichotomy, the key difference is the autonomy workers in more market-like platform business models have in setting prices (and also more room for negotiations with the client) compared to more hierarchical platform business models where the platform sets the rates.

Health and safety issues in platform work generally do not differ from comparable situations in the traditional labor market. However, responsibilities for ensuring regulated standards are often unclear, which in practice means self-responsibility of the worker. This is a particular problem in platform-determined work where the pay-by-task mechanism, the younger age—and hence inexperience and unfamiliarity of the worker—and the physical work environment pose more risks for the workers than in other types of platform work. In contrast, workers in the worker-initiated or client-determined type of platform work tend to be well familiar with health

Figure 9.3 Potential working conditions effects of different type of platform work [2].

and safety standards and are used to implement precautionary measures. From a policy perspective, health and safety becomes a concern if in more hierarchy-based platform business models responsibilities are unclear or unduly outsourced to the workers.

9.4 Conclusions and Policy Pointers

Platform work is a comparatively young phenomenon in Europe. It can be considered as a marginal employment form for the time being, but has been observed to dynamically develop, both in scale and scope, during the last decade. Accordingly, future growth is expected triggering intensive discussions on how to ensure that the economy and the labor market can benefit from the opportunities while minimizing the risks associated to platform work. In this context, it is strongly recommended to acknowledge the increasing heterogeneity within platform work, resulting in different advantages and disadvantages by type, which, in turn, require different policy approaches. The above discussions clearly show that a simple categorization by skills level of tasks or format of service provision (online vs on-location) is not sufficient to capture the complexity of platform work as regard its impact on labor market and working conditions. A combination of various elements determines the labor market and working conditions effects of this employment form and business model, and more nuanced analysis and policy considerations are required.

One of the current big discussion topics related to platforms is the employment status of workers. Across Europe, there does not exist any regulation clearly stating whether platform workers are employees or self-employed. The above analysis hints toward the strong influence of the business model of the platforms in assessing the factual classification of workers. One policy approach to tackle the issue could be to establish a default classification of employee/self-employed differentiated by platform work, with the requirement toward the platform to provide justification for a different than the default employment status.

A specific characteristic of platform work is algorithmic management and control as well as the importance of ratings for workers' access to tasks and working conditions. While they offer the opportunity of neutrality in decision-making, a common lack of transparency of the underlying logic and of portability of data across platforms, raises concerns regarding the fairness of the system and workers' access to redress if they feel unfairly treated. Transparency obligations for platforms as well as "ombudsman-like" services to assist workers could be considered to tackle this challenge.

Decent pay for globally competing platform workers as well as predictability of earnings and work schedules in some types of platform work should be ensured. Recently proposed EU directives could cover such expectations; their implementation and effectiveness should be explored. A challenge in this context is the above-mentioned unclear employment status as most of the protective EU directives target employees, only. Health and safety issues in platform work currently receive comparatively little attention and could be further reviewed.

Awareness raising and information provision toward workers and clients are recommended, as is focusing not only on physical health but also on psychosocial effects of algorithmic management.

References

[1] Codagnone, C., Karatzogianni, A. and Matthews, J. (2019). "Platform Economics. Rhetoric and Reality in the "Sharing Economy". London: Emerald Publishing.

[2] Eurofound (2019). "Platform Work: Maximising the Potential while Safeguarding Standards?" Luxembourg: Publications Office of the European Union.

[3] Hagiu, A. and Wright, J. (2015a). "Enabling versus controlling". Harvard Business School Strategy Unit Working Paper No. 16-002. Retrieved from http://ssrn.com/abstract=2627843 [accessed 20 November 2015].

[4] Hagiu, A. and Wright, J. (2015b). "Marketplace or Reseller?" Management Science, 61(1), 184-203. doi:10.1287/mnsc.2014.2042.

[5] Hagiu, A. and Wright, J. (2015c). "Multi-sided Platforms". International Journal of Industrial Organization. doi:10.1016/j.ijindorg.2015.03.003

[6] Evans, D. (2003a). "The Antitrust Economics of Multi-Sided Platform Markets". Yale Journal on Regulation, 20(2), 325-381.

[7] Evans, D. and Noel, M. (2005). "Defining Antitrust Markets When Firms Operate Two-Sided Platforms". Columbia Business Law Review, 3, 667-702.

[8] Evans, D. and Schmalensee, R. (2007). "The Industrial Organization of Markets with Two-Sided Platforms". Competition Policy International, 3(1), 151-179.

[9] Wright, J. (2004). "One-sided Logic in Two-sided Markets". Review of Network Economics, 3(1), 44-64.

[10] Cullen, Z. and Farronato, C. (2015). "Outsourcing Tasks Online: Matching Supply and Demand on Peer-to-Peer Internet Platforms". Standford: Department of Economics, Stanford University. Retrieved from http://web.stanford.edu/~chiaraf/matching_p2p_latest.pdf.

[11] Farronato, C. and Fradkin, A. (2015). "Market Structure with the Entry of Peer-to-Peer Platforms: The Case of Hotels and Airbnb". Standford University, in progress.

[12] Fradkin, A. (2014). "Search Frictions and the Design of Online Marketplaces". Retrieved from http://andreyfradkin.com/assets/Fradkin_JMP_Sep2014.pdf

[13] Fradkin, A., Grewal, E., Holtz, D. and Pearson, M. (2015). "Bias and Reciprocity in Online Reviews: Evidence From Field Experiments on Airbnb". EC'15: Proceedings of the Sixteenth ACM Conference on Economics and Computation, June 2015. https://doi.org/10.1145/2764468.2764528.

[14] Hagiu, A. and Wright, J. (2013). "Do You Really Want to Be an eBay?" Harvard Business Review, 91(3), 102-108.

[15] Horton, J. (2014). "Misdirected search effort in a matching market: causes, consequences and a partial solution". Paper presented at the Proceedings of the Fifteenth ACM Conference on Economics and Computation, Palo Alto, California, USA.

[16] Horton, J. and Golden, J. (2015). "Reputation Inflation: Evidence from an Online Labor Market". Retrieved from http://econweb.tamu.edu/common/files/workshops/Theory%20and%20Experimental%20Economics/2015_3_5_John_Horton.pdf.

[17] Gawer, A. and Cusumano, M. (2002). "Platform Leadership". Boston, MA: Harvard Business School Press.

[18] Caillaud, B. and Jullien, B. (2003). "Chicken & Egg: Competition among Intermediation Service Providers". The RAND Journal of Economics, 34(2), 309-328.

[19] Eisenmann, T., Parker, G. and Van Alstyne, M. (2006). "Strategies for Two Sided Markets". Harvard Business Review, 84(10), 92-101.

[20] Evans, D. (2003b). "Some Empirical Aspects of Multi-sided Platform Industries". Review of Network Economics, 2(3), 2194-5993. doi:10.2202/1446-9022.1026.

[21] Evans, D. (2008a). "Competition and Regulatory Policy for Multi-Sided Platforms with Applications to the Web Economy". Concurrences, 2, 57-62.

[22] Evans, D. (2008b). "The Economics of the Online Advertising Industry". Review of Network Economics, 7(3), 351-391. doi:10.2202/1446-9022.1154.

[23] Evans, D. (2009). "The Online Advertising Industry: Economics, Evolution, and Privacy". Journal of Economic Perspectives, 23(3), 37-60. doi:10.1257/jep.23.3.37.

[24] Evans, D. (2011). "Platform Economics: Essays on Multi-Sided Businesses". Competition Policy International. Available at SSRN: https://ssrn.com/abstract=1974020.

[25] Parker, G. and Van Alstyne, M. (2000). "Internetwork externalities and free information goods". Paper presented at the Proceedings of the 2nd ACM Conference on Electronic Commerce, Minneapolis, Minnesota, USA.

[26] Parker, G. G. and van Alstyne, M. W. (2005). "Two-Sided Network Effects: A Theory of Information Product Design". Management Science, 51(10), 1494-1504. doi:10.2307/20110438.

[27] Rysman, M. (2009). "The Economics of Two-Sided Markets". Journal of Economic Perspectives, 23(3), 125-143.

[28] Rochet, J.-C. and Tirole, J. (2003). "Platform Competition in Two-sided Markets". Journal of the European Economic Association, 1(4), 990-1029. doi:10.1162/154247603322493212.

[29] Rochet, J.-C. and Tirole, J. (2006). "Two-sided Markets: A Progress Report". RAND Journal of Economics, 37(3), 645-667.

[30] Filistrucchi, L., Geradin, D. and Van Damme, E. (2013). "Identifying Two-Sided Markets". World Competition: Law & Economics Review, 36(1), 33-60.

[31] Filistrucchi, L., Geradin, D., van Damme, E. and Affeldt, P. (2014). "Market Definition in Two-Sided Markets: Theory and Practices". Journal of Competition Law and Economics, 10(2), 293-339. doi:10.1093/joclec/nhu007.

[32] Armstrong, M. (2006). "Competition in Two-sided Markets". RAND Journal of Economics, 37(3), 668-691. doi:10.1111/j.1756-2171.2006.tb00037.x.

[33] Horton, J. (2010). "Online Labor Markets". Internet and Network Economics, 6484, 515-522. doi:10.1007/978-3-642-17572-5_45.

[34] Musthag, M. and Ganesan, D. (2013. "Labor dynamics in a mobile micro-task market". Paper presented at the Proceedings of the SIGCHI Conference on Human Factors in Computing Systems, Paris, France.

[35] Thebault-Spieker, J., Terveen, L. and Hecht, B. (2015). "Avoiding the south side and the suburbs: the geography of mobile crowdsourcing markets". Paper presented at the Proceedings of the 18th ACM Conference on Computer Supported Cooperative Work & Social Computing, Vancouver, BC, Canada.

[36] Teodoro, R., Ozturk, P., Naaman, M., Mason, W. and Lindqvist, J. (2014). "The motivations and experiences of the on-demand mobile workforce". In Proceedings of the 17th ACM Conference on Computer Supported Cooperative Work & Social Computing. Baltimore, Maryland, USA (pp. 236-247).

[37] Autor, D. (2008). "The Economics of Labor Market Intermediation: An Analytic Framework". Cambridge, Massachusetts: National Bureau of Economic Research.

[38] Autor, D. (2013). "The 'task approach' to labor markets. An overview". NBER Working Paper No. 18711. doi: 10.3386/w18711.

[39] Autor, D., Levy, F. and Murnane, R. (2003). "The Skill Content of Recent Technological Change: An Empirical Exploration". The Quarterly Journal of Economics, 118(4), 1279-1333.

[40] Autor, D. H. and Dorn, D. (2013). "The Growth of Low-Skill Service Jobs and the Polarization of the US Labor Market". American Economic Review, 103(5), 1553-1597. doi:10.1257/aer.103.5.1553.

[41] Autor, D. H., Katz, L. F. and Kearney, M. S. (2006). "The Polarization of the U.S. Labor Market". American Economic Review, 96(2), 189-194. doi:10.1257/000282806777212620.

[42] Dettling, L. (2016). "Broadband in the Labor Market: The Impact of Residential High Speed Internet on Married Women's Labor Force Participation". Industrial and Labor Relations Review. doi:10.1177/0019793916644721. http://ilr.sagepub.com/content/early/2016/04/20/0019793916644721.abstract

[43] Raja, S., Imaizumi, S., Kelly, T. and Paradi-Guilford, C. (2013). "Connecting to Work. How Information and Communication Technologies could help Expand Employment Opportunities". Washington DC: World Bank. Retrieved from: http://www-wds.worldbank.org/external/default/WDSContentServer/WDSP/IB/2013/09/09/000456286_20130909094536/Rendered/PDF/809770WP0Conne00Box379814B00PUBLIC0.pdf (23-10-2015).

[44] Rossotto, C., Kuek, S. and Paradi-Guilford, C. (2012). "New Frontiers and Opportunities in Work". Washington DC: World Bank. Retrieved from: http://www-wds.worldbank.org/external/default/WDSContentServer/WDSP/IB/2013/06/07/000356161_20130607151348/Rendered/PDF/782660BRI0P1280Box0377336B00PUBLIC0.pdf (23-10-2015).

[45] Agrawal, A., Lacetera, N. and Lyons, E. (2013b). "Does information help or hinder job applicants from less developed countries in online markets?" NBER Working Paper Series, 18720. doi:10.3386/w18720.

[46] Beerepoot, N. and Lambregts, B. (2015). "Competition in Online Job Marketplaces: Towards a Global Labour Market for Outsourcing Services?" Global Networks, 15(2), 236-255. doi:10.1111/glob.12051

[47] Galperin, H., Viecens, M. and Greppi, C. (2015). "Discrimination in Online Contracting: Evidence from Latin America". Ottawa: International Development Research Centre and the Canadian International.

[48] Ghani, E., Kerr, W. and Stanton, C. (2014). "Diasporas and Outsourcing: Evidence from oDesk and India". Management Science, 60(7), 1677-1698.

[49] Hong, Y. and Pavlou, P. (2014). "Is the World Truly 'Flat'? Empirical Evidence from Online Labor Markets". Fox School of Business Research Paper No. 15-045. Retrieved from SSRN: http://ssrn.com/abstract=2371748 (2-12-2015).

[50] Lehdonvirta, V., Barnard, H., Graham, M. and Hjorth, I. (2014). "Online labour markets – leveling the playing field for international service

markets?" Paper presented at the IPP2014: Crowdsourcing for Politics and Policy, Oxford, UK. Retrieved from: http://ipp.oii.ox.ac.uk/sites/ipp/files/documents/IPP2014_Lehdonvirta_0.pdf (15-10-2015).

[51] Mill, R. (2011). "Hiring and learning in online global labor markets". IDEAS Working Paper Series from RePEc.

[52] Ipeirotis, P. (2010a). "Analyzing the Amazon Mechanical Turk Marketplace". XRDS, 17(2), 16-21. doi:10.1145/1869086.1869094

[53] Ipeirotis, P. (2010b). "Demographics of Mechanical Turk". Department of Information, Operation, and Management Sciences, New York University. Retrieved from https://archive.nyu.edu/bitstream/2451/29585/2/CeDER-10-01.pdf [accessed 4 December 2015].

[54] Agrawal, A., Horton, J., Lacetera, N. and Lyons, E. (2013a). "Digitization and the contract labor market: A research agenda". NBER Working Paper Series, 19525.

[55] Florisson, R. and Mandl, I. (2018). "Platform Work: Types and Implications for Work and Employment – Literature Review". Dublin: Eurofound.

[56] Eurofound (2018). "Employment and Working Conditions of Selected Types of Platform Work". Luxembourg: Publications Office of the European Union.

[57] Fabo, B., Beblavý, M., Kilhoffer, Z. and K. Lenaerts (2017). "An Overview of European Platforms: Scope and Business Models". Luxembourg: Joint Research Centre, Publications Office of the European Union.

[58] De Groen, W. P., Lenaerts, K., Bosc, R. and Paquier F. (2017). "Impact of Digitalisation and the On-demand Economy on Labour Markets and the Consequences for Employment and Industrial Relations". Brussels: European Economic and Social Committee.

10

Workplace Innovation and Industry 4.0: Creating Synergies between Human and Digital Potential

Peter Totterdill

Workplace Innovation Europe, Cork, Ireland, and Glasgow, UK

Abstract

From a critical perspective, Industry 4.0 is at risk of being no more than the latest in the long line of technological predictions based on exaggerated claims. It runs the risk of drawing corporate decision-makers into patterns of investment that ultimately fail because they ignore the importance of synergy between the design and implementation of technologies on the one hand, and human and organizational factors on the other. There is a need to articulate the choices and alternative narratives surrounding Industry 4.0. Yet the technological advances represented by Industry 4.0 potentially offer real economic and also social benefits. At the same time, realizing this potential and avoiding the mistakes of the past means recognizing the importance of a new and more inclusive paradigm of innovation. The challenge is that of reconciling the ordered, rational organization of work offered by emergent technologies with creative, dialogical, serendipitous, and even chaotic human interactions that can stimulate innovation.

Keywords: Competitiveness, determinism, digitalization, employment, empowerment, Industry 4.0, innovation, quality of working life, skills, technology, workplace innovation

10.1 Introduction: A Critical Perspective

When does a popular idea begin to outlive its usefulness, gradually obscuring the reality it was intended to explain? And how do we distinguish transient fashions in thinking and practice from underlying truths?

In recent decades, we have seen and sometimes contributed to successive yet always short-lived fashions in predicting the future of work and the economy: flexible specialization [1]; the virtual organization [2]; the end of work [3]; the new economy (1990s); sociocracy and the death of hierarchy (e.g., [4]); the millennial workforce, and so on. "The Fourth Industrial Revolution," more commonly known in Europe as "Industry 4.0" since its adoption by the German Federal Government, is the latest in this long line of attempts to make sense of emerging forces in what is undoubtedly an increasingly volatile global economic environment.

In line with its predecessors in prediction, Industry 4.0 contributes important insights and enhances understanding of the challenges and opportunities facing corporations and policymakers alike (Fig. 10.1). Yet claims that it offers a comprehensive, global narrative on the future of work and the economy, and indeed that it represents an inevitable as well as a desirable development should be treated with caution. The concept is being driven "by computer scientists, engineers, innovation policy actors, influential business associations, and larger technology-intensive enterprises" ([5]), and as with other fads, it contains much speculation, contradictory evidence and, most importantly, a tendency to conceal choices. As the German researchers Kopp et al. argue [6], Industry 4.0 can easily become "Technological Determinism 4.0," repeating the mistakes of previous eras in which technocratic reductionism became so pervasive in some industrial settings leading to an expensive failure. Corporate and public discourse needs to recognize the existence of alternative narratives and competing choices.

Kopp et al. [6] suggest that when the initial, still undiminished euphoria surrounding Industry 4.0 dies down, the choices and dilemmas, which surround it will become much more evident. Indeed economic and workplace futures in democratic societies will continue to be shaped by choices and decisions made by diverse stakeholders including politicians, scientists, thinkers, and individuals rather than determined by a linear technological imperative.

In Europe, the public policy approach centers on raising the competitiveness of advanced manufacturing through enhanced innovative capacity, productivity, growth, and employment, recognizing the critical role

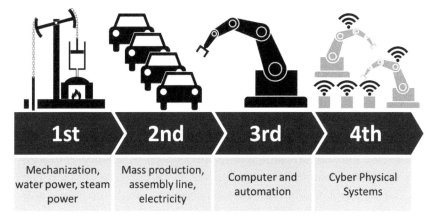

Figure 10.1   Industry 4.0. Source: Schwab, 2016 [9].

of human factors and "inclusive growth" [7, 8]. This chapter explores the corporate choices and opportunities involved in realizing that goal.

10.2 The Promise (and Threat) of Industry 4.0

"Industry 4.0" describes a new level of organization and management of the entire value chain across the product life cycle, able to meet increasingly individualized customer wishes so that even one-off items can be manufactured profitably. It can form extended value creation chains linking manufacturers with their suppliers and customers, encompassing idea generation, product development, production, delivery to the end customer, and eventually recycling.

The European Parliamentary Research Service [10] summarizes the new and innovative technological developments on which Industry 4.0 depends as follows:

- The application of information and communication technology (ICT) to digitize information and integrate systems at all stages of product creation and use (including logistics and supply), both inside companies and across company boundaries.
- Cyber-physical systems that use ICTs to monitor and control physical processes and systems.
- Network communications including wireless and internet technologies that link machines, work products, systems, and people.

- Simulation, modeling and virtualization in the design of products and the establishment of manufacturing processes.
- Collection of vast quantities of data, and their analysis and exploitation, either immediately on the factory floor, or through big data analysis and cloud computing.
- Greater ICT-based support for human workers, including robots, augmented reality and intelligent tools.

Transmission of data through the manufacturing chain, automation of production, and the use of configurable robots lead to greatly enhanced flexibility and mass customization since a variety of different products can be produced in small batches in the same facility. Such flexibility also encourages innovation since prototypes or new products can be produced quickly without complicated retooling or setting up new production lines. Digital designs and virtual modeling of manufacturing process can reduce the time between product conception and delivery.

Customers will be able to be more involved in the design process. Production can also be located close to the customer because, if manufacturing is largely automated, it does not need to be "off-shored" or located in low labor cost countries, and "re-shoring" is already occurring in parts of Europe[1].

Integrating product development with digital and physical production has also been associated with large improvements in product quality and significantly reduced error rates since data from sensors can be used to monitor every piece produced rather than using sampling to detect errors, and error-correcting machinery can adjust production processes in real time.

Productivity can also increase. By using advanced analytics in predictive maintenance programs, manufacturing companies can avoid machine failures on the factory floor and cut downtime significantly. Some companies are already setting up "lights out" factories where automated robots continue production without light or heat after staff leave for home.

Industry 4.0 can also enable long-sought changes in business models. Rather than "low road" competitive strategies based primarily on cost, Industry 4.0 may allow companies in high labor cost countries to compete on the basis of innovation being able to deliver new products rapidly to customer-driven designs with the assurance of high quality standards. Falling costs for digital technologies may also help to close the productivity gap

---

[1]https://reshoring.eurofound.europa.eu/

between small- and medium-sized enterprises (SMEs) and large companies found in some European countries. Even more significantly, technologies such as three-dimensional (3D) printing have the potential to decentralize the production of many consumer goods to local or even domestic sites, while current corporate manufacturers become pure software companies[2].

Some argue[3] that, in addition, Industry 4.0 will address and solve social and environment challenges such as resource efficiency and demographic change. For example, workers can be released from routine tasks, enabling them to focus on creative, value-added activities. Older workers will be able to extend working lives and remain productive for longer, ameliorating the impact of an ageing workforce in several European countries. Flexible work organization should also enable workers to combine work, private lives, and continuing professional development more effectively.

Yet these promises conceal considerable anxiety about how the transition to a better, brighter future will affect current jobs and businesses. Polarization in European labor markets has been observed for some time (e.g., Ref. [11]). Low-skilled workers are offered few opportunities to upgrade their skills while those with higher education are offered more. The OECD Jobs Study [12] showed that this process had already begun in the mid-1980s, and it has continued ever since, including Europe. Lundvall et al. [13] found that the growing income inequality excluded and marginalized low skilled workers from new employment opportunities. Arguably the consequences of this widening and cumulative process of marginalization are reflected in growing political volatility represented by the election of President Trump in the United States, Brexit in the United Kingdom, and the rise of the far right in countries such as Germany, Hungary, and Poland.

It is certainly clear that that greater use of digital industrial technologies will reduce the number of traditional assembly and production jobs, yet the scale of the loss is heavily contested. Based on a detailed analysis of several forecasts and projections, Bakhshi et al. [14] show that alarming and widely-publicized headline findings suggesting, e.g., that "47% of US workers' jobs are at high risk of automation" have been challenged by other researchers, and that once detailed task variations are taken into account the figure maybe closer to 9%. In the United Kingdom, they predict that by 2030 "Around one-fifth (of employees) are in occupations that will likely shrink," and that these

---

[2]www.forbes.com/sites/ricksmith/2015/07/07/5-incredible-trends-that-will-shape-our-3d-printed-future/#58799301fa48
[3]See for example www.workplaceinnovation.org/nl/kennis/kennisbank/industry-4-0/1241

are mainly in low- or medium-skilled occupations in both manufacturing and administration. These projections are much lower than other recent studies of automation have suggested, reinforcing a view that in many occupations complete automation is not realistic and that improvements in productivity will be achieved mainly through enhancing human labor through digital assistance rather than replacing it[4].

On the positive side, Gregory et al. [15] estimate that automation boosted net labor demand across Europe by up to 11.6 million jobs over the period 1990–2010. Its job-destroying effects were offset by lower unit costs and prices which stimulate higher demand for products, and that surplus income from innovation was converted into additional spending, so generating demand for extra jobs in more automation resistant sectors.

Bakhshi et al. [14] predict that in the United Kingdom "around one-tenth of the workforce are in occupations that are likely to grow as a percentage of the workforce" by 2030. Creative, digital, design, and engineering occupations have bright outlooks and are strongly complemented by digital technology. They also cite US data, which suggest that roles such as management analysts and training, development and labor relations specialists, and all occupations associated with the reorganization of work, are projected to grow. However, "roughly seven in ten people are currently in jobs where we simply cannot know for certain what will happen."

10.3 Old Skills for New Jobs

A common feature of projections about the employment impact of Industry 4.0 lies in the prediction that higher order cognitive skills will feature prominently in the future demand for labor. Originality, fluency of ideas, and active learning will be highly important as well as system thinking, judgment, and decision-making skills, not just because they are necessary to manage complex technological systems but also because they feed the creativity required by a culture of innovation.

Social skills will also continue to grow in importance in building customer service and negotiating the coordination frameworks required by Industry 4.0, which will often involve the creation of high-trust relationships across the globe [14, 16, 17]. Strikingly, nearly all US job growth since 1980

---

[4]See for example www.forbes.com/sites/haroldsirkin/2016/04/19/advanced-manufacturing-is-not-a-job-killer-its-a-job-creator/#220cfa9d5ddd

has been in relatively social skill-intensive occupations, and occupations with high analytical but low social skill requirements shrank over the same period [18].

Bakhshi et al. [14] express optimism that "occupation redesign coupled with workforce retraining" could promote growth in occupations whose future is uncertain and enable the adaptation of workers whose jobs are under threat. Conceivably, digitally assisted work environments could ease the transition to new jobs and even encourage some older workers to return to work.

Lundvall's emphasis on the importance of "discretionary learning jobs" [19] is helpful in this context. Discretionary learning refers to a job situation where the employee has a certain freedom (discretion) to decide how to solve problems and where, in consequence, (s)he continuously learns new skills. It stands in contrast to Taylorist work where there is both little freedom to act and very limited learning for the employee. Arundel et al. [20] found very clear patterns showing that in countries where ordinary workers are engaged in discretionary learning jobs, domestic enterprises were more engaged in radical innovation. Yet overall, less than 40% of Europe's workers are employed in discretionary learning jobs [19].

The challenge remains. Europe's track record in managing the transition of workers in declining industries, to secure skilled employment in other fields, is at best patchy—and it is impossible not to think of the continuing marginalization or exclusion of former coal miners, steel workers, or sewing machinists in many communities. Active interventions to support workforce adjustment are certainly possible but this is no guarantee that this will happen, as Lundvall et al. [13] found in their analysis of labor market polarization discussed above.

10.4 The Emergence of a New Innovation Paradigm

As we have seen, the Industry 4.0 narrative emphasizes its potential to facilitate product and service innovation through digital design, virtual modeling, and rapid prototyping. The key challenge is to understand the organizational conditions under which human creativity can realize this potential.

Innovation has often been seen as the prerogative of a scientific, entrepreneurial, or management élite, yet recent research shows that it thrives in egalitarian learning economies where ordinary workers enjoy jobs that make full use of their skills and learning capacity (e.g., [13]). Likewise the traditional view of innovation has been challenged from several other complementary directions, e.g., "open innovation" [21], "customer-driven innovation" [22], "cocreation" [23], and "networked innovation" [24] mirror important aspects of an emerging innovation paradigm that has to be considered alongside the technological dimensions of Industry 4.0. German "Learning Factories" [25] are a concrete example of the ways in which innovation, learning, and work can be integrated in the workplace. Tangible examples can also be found in the rise of "FabLabs" and the "Maker Movement"[5]. These have close links to "free and open source" thinking including the open source software movement, sharing the philosophy that all can be empowered to use and shape creative technologies. They are being created by universities and colleges, by not-for-profit entities in local communities and, increasingly, by companies who want to supercharge innovation by forming spaces where frontline employees, customers, and other stakeholders can think "out of the box," collaborate, and discover the potential for serendipitous breakthroughs.

Totterdill et al. [26] argue that in the early 1990s a significant shift in Europe's economy could be observed, fueled by information technology. This shift reversed the historical pattern in which tangible capital was considered the main asset in companies. From around 1990, investments in intangible capital (as a percentage of adjusted GNP) such as patents, R&D, marketing, and organizational competences became higher than investments in tangible capital [27]. The conviction grew in Europe that "social innovation" in the workplace could be more important than "technological innovation" in explaining company performance [28]. Developing and utilizing the full range of skills and competences in the present and future workforce is therefore a vital component of competitive and knowledge based global economy [29]. Likewise, the *OECD Innovation Strategy*, the culmination of a 3-year, multidisciplinary and multistakeholder effort, emphasizes that "empowering people to innovate" and "fostering innovative workplaces" is important for creativity, innovation, and productivity [30]. Moreover, organizations only achieve a full return on investment in technological innovation if it is embedded in workplace innovation, in other words making the technology work by achieving a full synergy with human and organizational factors.

---

[5]See for example www.create-hub.com/comment/the-maker-movement-shifting-uk-manufacturing/

Jensen et al. [31] used survey data from around 700 Danish firms to link their mode of learning to innovation performance, and the statistical analysis led to four clusters of firms: low learning, science-based learning, experience-based learning, and a combination of science- and experience-based learning. Science-based learning refers to a process where systematic research plays a major role, and the knowledge produced is often codified. Experience-based learning refers to learning by doing, learning by using, and learning by interacting, and here much of the knowledge remains tacit, embodied in people, and embedded in organizations.

Jensen et al. [31] show that firms engaged in innovation need to combine the two modes. While firms that practiced one of the two learning modes were twice as innovative as those with low learning, firms that combined the two modes were five times as innovative as those with low learning. Innovation management at corporate level therefore needs to focus on building a learning organization and a pervasive culture of "high involvement" and "employee-driven" innovation [32, 33].

10.5 High Involvement Innovation and Industry 4.0

To summarize the argument so far, the potential of Industry 4.0 will only be fully realized if the technocratic reductionism of previous eras is rejected, and there is a reconciliation of what might be seen as two conflicting models [34]; one focused on structure and order to attain the rational organization of work, and the other in which creativity and human dialogue drive innovation (Fig. 10.2).

In reconciling these two models, our starting point lies with the vast and growing body of evidence demonstrating that workplace practices which empower employees to contribute ideas and be heard at the most senior levels of an organization lead to improved productivity and capacity for innovation, as well as enhanced workforce health and engagement [33, 35, 36]. Such practices have increasingly been described as "workplace innovation" since the early years of the present century.

According to the Hi-Res study, a meta-analysis of 120 case studies across 10 European countries, workplace innovation takes diverse forms but is always characterized by:

"... a clear focus on those factors in the work environment which determine the extent to which employees can develop and use their competencies and creative potential to the fullest extent, thereby enhancing

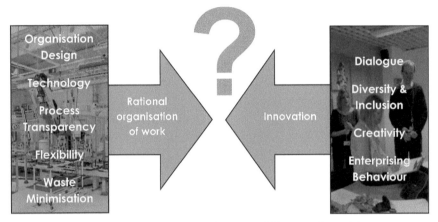

Figure 10.2   Structure and order versus dialogue and creativity.

the company's capacity for innovation and competitiveness while enhancing quality of working life." [37].

Such factors in the work environment include empowering job design, self-organized teamworking, structured opportunities for reflection, learning, and improvement, high involvement innovation practices, the encouragement of entrepreneurial behavior at all levels of the organization, and employee representation in strategic decision-making.

Workplace innovation is an inherently social process, creating self-sustaining processes of development by learning from diverse sources and by experimentation. It seeks to build bridges between the strategic knowledge of the leadership, the professional and tacit knowledge of frontline employees, and the organizational design knowledge of experts, engaging all stakeholders in dialogue in which the force of the better argument prevails [38, 39].

Thus, in defining workplace innovation, it is important to recognize both process and outcomes. The term describes the participatory *process* of innovation, which leads to *outcomes* in the form of participatory workplace practices. Such practices grounded in continuing reflection, learning, and improvement sustain the process of innovation in management, work organization, and the deployment of technologies.

10.6 Defining Workplace Innovation

Workplace innovation now occupies an important place in EU innovation and competitiveness policy. It has led to the creation of the European

Commission's Workplace Innovation Network[6] (EUWIN) in 2012, jointly led by the TNO[7] and Workplace Innovation Europe[8].

The creation of the EUWIN provided an opportunity to address the need for a new type of dialogue between researchers and practitioners. EUWIN's task is to promote the dissemination of workplace innovation throughout Europe through knowledge sharing and dialogue[9].

With limited resources, a clear framework for communication was a priority for the EUWIN partners. Workplace innovation is a hard-to-grasp concept, and it was important to make it more communicable, without breaking the link with the large and complex body of research evidence that underpins it. This led to the formulation of *The Essential Fifth Element* concept (Fig. 10.3) by the Workplace Innovation Europe team as a means of providing practical and actionable insights into evidence and experience underpinning workplace practices associated with high performance, innovation, and quality of working life [33].

The Essential Fifth Element is based on an analysis of more than 100 articles and a similar number of case studies from which four main bundles of workplace practices (or "elements") were detected, each associated with improved performance and quality of working life:

1. Jobs and teams
2. Organizational structures, management, and procedures
3. Employee-driven improvement and innovation
4. Cocreated leadership and employee voice

Each of these bundles does not exist in isolation but is influenced, for better or worse, by the others. Workplace innovation cannot be reduced to fragmented practices if it is to realize its potential. The literature emphasizes the importance of internally consistent policies and practices combining different forms of representative and direct participation in achieving superior outcomes for organizations and their employees, which are greater than the sum of individual measures [40, 41, 42]. *The Essential Fifth Element* can also be related to the "configurational approach of strategic human resource management" (SHRM): "In general, configurational theories are

[6]http://ec.europa.eu/growth/industry/innovation/policy/workplace/index_en.htm
[7]www.tno.nl
[8]www.workplaceinnovation.eu
[9]http://uk.ukwon.eu/euwin-resources-new

Figure 10.3   The Essential Fifth Element: conceptualizing the characteristics and outcomes of workplace innovation.

concerned with how the pattern of multiple independent variables is related to a dependent variable rather than with how individual independent variables are related to the dependent variable" ([43]. Thus, bundles of practices are more effective than separate interventions [44]. Likewise, studies of failed workplace innovations emphasize the role of "partial change" in undermining the introduction of empowering working practices [45]. This provides the starting point for *The Essential Fifth Element*.

In medieval alchemy, the "fifth element" was created when the four essential elements (earth, wind, air, and water) that make up the world combine. As a metaphor, it provides a practical, actionable way of understanding the cumulative impact when empowering workplace practices are aligned at every level of the organization, embracing both direct and indirect participation. Such alignment can create a tangible effect in workplaces which is hard to quantify but which is often described in terms of "engagement" and "culture," and is strongly associated with convergence between high performance and high quality of working life [33, 36, 46, 47, 48, 49]. The concept is explained further on the EUWIN Knowledge Bank and in a short film[10]. *The Essential Fifth Element* has been disseminated widely in Europe, perhaps most systematically through government-funded programmes in Scotland [50].

---

[10]https://workplaceinnovation.eu/this-is-workplace-innovation/

10.7 Workplace Innovation and Digital Technologies: Creating Synergies in Practice

Organizations have choices in the way that they implement new technologies. Technologies can be used to remove or deskill jobs ("digital Taylorism") or to use and develop workforce skills more effectively by making jobs more complex, challenging, and intrinsically satisfying [51, 52]. The question then becomes one of informing and shaping that decision process: less "what can the technology do?" and more "workplace innovation enables us to build a high performing organization: how can technology support it?"

Workplace innovation is defined by specific bundles of working practices that are strongly associated with the simultaneous realization of high performance and high-quality working lives. The section explores how these bundles as defined by *The Essential Fifth Element* provide the framework for technological investment fully aligned with the organizational strategy and potential.

10.7.1 The First Element: Jobs and Teams

Building workplaces in which employees can develop and deploy their competencies and creative potential begins with job design. Well-designed jobs that provide constructive challenges, opportunities for day-to-day problem solving, variety, and collaboration help people manage the demands placed on them and avoid the psychological stress and disengagement associated with repetitive and disempowering work [53, 54, 55, 56]. Moreover, through exercising discretion in such "complex jobs" employees acquire skills that are transferable, increasing their adaptability and resilience within the organization and their employability outside it, even in quite different occupations [57].

In De Sitter's sociotechnical systems design (STSD) theory, the central idea is the balance between "control requirements" (quantitative and qualitative demands) and "control capacity" (job control). "It's not the problems and disturbances in the work that cause stress, but the hindrances to solve them" [58]. To maintain this balance, control capacity is required regarding the performance of a given job on individual job level as well as regarding the division of labor on production group and plant level: "from complex organizations with simple jobs to simple organizations with complex jobs" [59]. Besides internal control capacity, complex jobs also include participation in external control activities at production group and plant level, e.g., shop floor consultation on processes, division of labor, and targets. The aim of such sociotechnical design is to simultaneously result in improved organizational performance, quality of working life, and better labor relations.

Of course, robotic technologies can play a role in removing repetitive, physically demanding, low-skill work in ways that enable employees to focus on higher value tasks. We are aware of cases, such as the Danish manufacturing company BM Silo, in which shop floor workers were trained in programming skills and given time to "play" with the robots to find ways of deploying them that enhanced their own productivity and job quality[11]. The development of cobots (collaborative robots responsive to worker movements and capable of learning) further enhances the potential for human-centered technological deployment [60]).

Job complexity and discretion can also be supported by remote assistance aids to support problem solving and decision-making without micromanagement or immediate supervision, whilst at the same time reducing quality and compliance risks. Such technologies include augmented reality headsets capable of "walking" employees through unforeseen or previously unencountered tasks together with 3-way calling from remote locations.

Effective job design must develop in synchrony with the wider organizational context. The key concept here is self-managed teamworking, one of the defining characteristics of workplace innovation, with deep roots in European thinking about management and organization dating back to the work of the Tavistock Institute in the 1940s and 1950s. Extensive research demonstrates that empowered and self-managed teams are more productive in factories and offices, provide better customer service, and even save lives in places like hospitals [37, 61, 62].

However, "teamwork" is increasingly used to describe such a diverse range of workplace situations that arguably the term is in danger of becoming meaningless. While teamworking may refer to a general "sense of community," or a limited enlargement of jobs to enhance organizational flexibility, empowered teamworking will involve a radical reappraisal of jobs, systems, and procedures throughout the whole organization [63]. "Real," self-managed teams give members responsibility, individually and collectively, for planning and scheduling their own work, for shared

---

[11]Presentation by Dorte Martinsen (MD, BM Silo), Workplace Innovation Masterclass, Stirling, 4th November 2019.

knowledge and problem solving, and for productive reflection and improvement [37, 61, 62].

Technologies support self-managed teamworking through the wide distribution of intelligence and information. Real-time data can be pushed across the organization, enabling and empowering team-based planning and scheduling as well as responsive decision-making on the job. Cornerstone, a Scottish organization employing 2000 social care workers, made a radical shift toward a flat structure using collaborative technologies, which allow self-managed team to balance workloads remotely, reflecting unpredictable client needs in real time.

Data flows also enable teams to see the impact of their own performance on other parts of the organization, and to inform learning and improvement.

10.7.2 The Second Element: Organizational Structures, Management, and Procedures

The Fifth Element recognizes the need for a consistent approach to empowerment, learning, and development running through every aspect of corporate policy from reward systems and performance appraisal to flexible working and budget devolution.

Hierarchical management layers inevitably put distance between decision-making and the frontline, disempowering and diminishing the voice of those at the lower levels as well as creating an implementation gap. Hierarchy breeds caution amongst managers, encouraging decisions to be delegated upward with the consequent loss of productivity and responsiveness. Vertically organized structures create silos and add to the difficulties of building bridges between functional specialisms. This often causes frustration in resolving day-to-day issues and can have a particularly negative effect on the capacity for innovation [64].

Flat organizations rely on a decentralized approach to management and require a high degree of employee involvement in decision-making [65]. Control in flat companies lies in mutual agreements between self-managing, self-organizing, and self-designing teams and employees who take personal responsibility for satisfactory outcomes. This in turn empowers employees, facilitates information sharing, breaks down divisions between roles, shares competencies, and uses organization-wide reward systems.

Automated alignment and presentation of data across the organization can support flatter structures by enabling horizontal decision-making between teams, better integrated product/service flow, faster problem solving, and freeing managers from involvement in routine planning and problem solving.

10.7.3 The Third Element: Employee-Driven Innovation and Improvement

Continuous improvement is now widely understood as a driver of quality and efficiency, and many organizations build a set of internal reflexive mechanisms designed to harness the day-to-day experiences of employees. Systematic opportunities for shared learning and "productive reflection" [66] are well embedded in these workplaces. Connected information systems can provide the foundations for continuous improvement, monitoring and distributing performance data horizontally across the business to identify and inform opportunities for change.

Beyond incremental improvements, studies of innovation emphasize the importance of large numbers of people empowered to act in entrepreneurial ways in pursuit of shared goals [67, 68]. This is reflected in times and spaces where employees can step back from their functional tasks to discuss ideas for new products, services, or processes openly and creatively with coworkers. Time-out sessions, hackathons, and FabLabs can become sources of constructive dialogue, creativity, and innovation[12], recognizing the importance of experimentation and "fast failure," and avoiding blame when things go wrong. Embedding such practices in company culture typically appears in recommendations for the successful implementation of digital technologies (see, e.g., Ref. [17]).

Some companies, arguing that new ideas can come from anyone, reject the idea of setting up a separate innovation team. Networks of volunteer "guerrillas," recruited from every level of the organization, trained in facilitation techniques, and empowered to ask difficult questions can be used to establish a pervasive culture of innovation[13].

Connected business and social communication systems, available to all employees irrespective of their location, stimulate creative thinking and enable collaborative innovation; these technologies can include ideation platforms, online forums, and distributed access to data.

---

[12] http://uk.ukwon.eu/learning-reflection-and-innovation-new
[13] See for example http://uk.ukwon.eu/met-office

10.7.4 The Fourth Element: Co-created Leadership and Employee Voice

Workplace innovation defines leadership as a creative and collective process [68], less concerned with the central, charismatic individual, and more with the creation of opportunities for employees to seize the initiative and contribute to decision-making. Such "shared and distributed leadership" relates to a concern with empowerment [68] and building organizational capability [70]; it is therefore a key element of workplace innovation in that it helps to release the full range of employee knowledge, skills, experience, and creativity [33].

Employee voice describes the alignment of strategic priorities and decision-making at senior levels with the practical knowledge, experience, and engagement of employees throughout the organization. It brings together *direct participation* through, e.g., self-managed teams and improvement groups, with *representative participation* in the form of employee or union-management partnership forums.

Representative participation, or workplace partnership between management, employees, and/or trade unions is an important aspect of this process of cocreation [71]. At its most basic level, partnership agreements and structures are a way of dealing proactively with industrial relations issues, ensuring early consultation on pay and conditions, employment changes and organizational restructuring. Representative partnership structures on their own may have little direct impact on performance or quality of working life. Rather they can exert a positive influence on the development of activities and practices that do so [72]. Workplace partnership thus moves away from its traditional focus on industrial relations, emerging as a potentially important driver of, and resource for, organizational innovation in the broadest sense [73, 74].

When partnership arrangements exist alongside the types of participative workplace practices described in the previous three elements it creates a system of mutually reinforcing practices leading to improved information sharing, greater levels of trust, reduced resistance to change, and heightened performance.

Technology can also help to close the gap between leaders and their employees. Integrated business and social communication systems accessible throughout the workforce enable real-time two-way interaction from the shop floor to the senior team, improving decision-making by bringing together the strategic perspectives of senior teams with the tacit knowledge of frontline workers. These technologies enable the challenges and opportunities facing senior teams to be addressed through early engagement and open dialogue with relevant employees, cocreating solutions rather than imposing them from the top.

10.8 Workplace Innovation as an Enabler of Digital Technologies

Booth Welsh, an engineering services company based in Scotland, is an example of an organization whose workplace innovation journey has led it into readiness for the adoption of digital technologies. The company recognized that engaging its workforce more effectively was central to future competitiveness. After joining a publicly funded workplace innovation development program [50], Booth Welsh surveyed its entire workforce using the Workplace Innovation Diagnostic® based on *The Essential Fifth Element*. The results showed that employees experienced few opportunities to take part in improvement and innovation, often worked in functional silos with little interaction with the rest of the business, and felt little connection with company strategy. A three-year program was created to address the root causes of these challenges (Table 10.1).

For Booth Welsh, the result has been a tangible unleashing of employee creativity and initiative, ranging from minor incremental improvements to strategically significant innovations. Critically it has introduced "Industry

Table 10.1 Booth Welsh's development program based on "The Essential Fifth Element".

Jobs and Teams	Decision-making is being devolved to self-managed teams, increasing frontline empowerment, and responsibility for achieving strategic targets.
Structures, Management, and Procedures	A flattened management structure is anticipated, enabling the organizational structure to be redesigned around workflow rather than specialist functions. Line managers will be refocused on activities relating to their individual strengths.
Employee-Driven Innovation and Improvement	Co-Labs bring people together across different functions to identify potential product, service, and process innovations. Dedicated innovation spaces have been created, accessible by all employees.
Cocreated Leadership and Employee Voice	All employees are encouraged to contribute to the development and implementation of four business-critical "strategic pillars," each comprising a series of working groups and actions.

4.0" as a narrative that is beginning to run throughout the business, leading to the acceptance and introduction of digital technologies both internally and in the services offered to clients.

10.9 People-Centered Change

Change is rarely a straightforward linear exercise. It usually involves experimentation, failure, and a willingness see failure as an opportunity for learning and development. It requires consistency of purpose combined with a willingness to rethink the vision and objectives set out at the start of the journey. The more you try to change an organization, the more you learn about it. Understanding of the nature and extent of the change required will deepen as the journey progresses.

While we need more evidence of the successful integration of digital technologies with human and organizational strengths, it is possible to draw some very broad guidelines for positive change (Table 10.2). Above all, it means making change happen with people, not to people.

Table 10.2   People-Centered Change.

CHANGE WITH PEOPLE
Embed employee-driven innovation and improvement in the DNA of the organization Establishing regular opportunities for employees at all levels to think creatively, to contribute their ideas, to experiment, and to work collaboratively with others on ideas for innovation and improvement creates a workforce culture receptive to digital innovation and enhances readiness for change.
Be clear about what you want to achieve Digital technologies are only a means to an end and may not be the best way of achieving it. Their introduction must be driven by, and fully aligned with, the organization's strategic goals and imperatives.
Start from a systemic view of the organization Incremental innovations run the risk of unintended consequences elsewhere or being undermined by interdependent practices in the organization. Creating a system of aligned and mutually reinforcing practices enhances outcomes and helps prevent innovation decay.
Interrogate and reveal the job design assumptions built-in to the technology The design of IT systems and production machinery, e.g., can steer organizations toward the creation of very specialized roles and functions, leading to workforce fragmentation and the narrowing of skills. On the other hand, technologies can be designed in ways that distribute information more widely, enhance the functional flexibility of individuals and teams, and build workforce capacity for problem solving, improvement, and innovation.

(Continued)

Table 10.2 Continued

Engage and empower operators and others whose work is affected by the technology in implementation

Day-to-day involvement in the production and delivery of a product or a service creates tacit knowledge about "what works," building on the cumulative learning and experience from both successes and failures. Operators are often the "organizational memory"; they acquire an innate capacity for creative problem solving and provide a valuable resource beyond that of technical manuals and standard operating procedures, both in day-to-day operations and during innovations. Engaging operators and stakeholders at every stage of implementation avoids errors that may remain undetected by technical experts alone, and may stimulate unexpected process innovations and improvements.

Upskill people

Technical skills involved in the implementation of digital technologies maybe acquired through both formal and informal means during the planning and implementation stages of digital innovation. Realizing the full synergy between human and digital potential also requires the development of wider business and self-efficacy skills, for example:

- Planning and scheduling to support delegated decision-making
- Communication skills
- Collaborative skills to support self-managed teamworking
- Creative thinking skills to support problem solving and employee-driven innovation
- Emotional intelligence and self-awareness

10.10 Conclusion

Industry 4.0 must be approached from a critical perspective, not least because of the hype and exaggerations, which surrounds its claims and potential impact. Drawing on past history, there are real dangers that a technocratic-driven narrative will draw corporate decision-makers into reductionist models and patterns of investment that ultimately fail because they ignore the importance of synergy between the design and implementation of technologies and human and organizational factors. There is a need for Europe-wide programs of experimentation and shared learning to create a better understanding of how synergies between human and digital potentials can be created in different sectoral and organizational contexts and indeed to understand that process of contextualization itself.

This chapter has discussed two separate narratives: the first focused on the ordered, rational organization of work offered by emergent technologies; the second on the creative, dialogical, serendipitous, and even chaotic human interactions that can stimulate innovation. Reconciling these narratives is essential if past mistakes are to be avoided and the positive potential of Industry 4.0 is to be realized.

The concept of workplace innovation, predating Industry 4.0 by a decade or so, prefigures many of its attributes. Like Industry 4.0, workplace innovation also seeks a transition between business models focused on cost-based competition to those based on innovation. It seeks the removal of monotonous work and its replacement with jobs focused on analysis, problem-solving, judgement, social interaction, and creativity.

Blending the ordered rationality of engineering and technology with the empowering and creative practices associated with workplace innovation will not be easy, and certainly challenges established cultures in many large corporate organizations. From our own experience of working with engineers and scientists, as well as leaders in advanced technology companies, resistance to change is a powerful force even where the business case is clear.

Predictably many corporate decision-makers will choose what they perceive to be safe, technocratic routes, which leave existing top-down or paternalistic cultures and working practices intact. Yet such risk-averse strategies ignore the lessons of previous eras, and indeed those of recent economic crises, which show that survival is not compulsory even for the largest players.

Acknowledgments

I gratefully acknowledge the considerable assistance provided by Booth Welsh, an integrated engineering services company based in Irvine, Scotland. Our collaboration with the Booth Welsh team has greatly enhanced our understanding of the potential for synergies between human and digital potential.

References

[1] Piore, M. J. and Sabel, C. (1984). "The Second Industrial Divide". New York: Basic Books.
[2] Davidow, W. and Malone, M. (1993). "The Virtual Corporation: Structuring and Revitalizing the Corporation for the 21st Century". New York: Harper.
[3] Rifkin, J. (1995). "The End of Work: The Decline of the Global Labor Force and the Dawn of the Post-market Era". New York: Putnam Publishing Group.

[4] Endenburg, G. (1998). "Sociocracy: The Organization of Decision-making".The Netherlands: Eburon.

[5] Hirsch-Kreinsen H. (2014). "Wandel von Produktionsarbeit – Industrie 4.0". In: WSI Mitteilungen 6/2014.

[6] Kopp R., Howaldt J. and Schultze J. (2016). "Why Industry 4.0 needs workplace innovation: a critical look at the German debate on advanced manufacturing". European Journal on Workplace Innovation, 2, 1, 7-24.

[7] European Commission (2010). "EUROPE 2020. A strategy for smart, sustainable and inclusive growth". COM (2010) 2020 Brussels, 3.3.2010.

[8] European Commission (2020). "Shaping Europe's Digital Future". Luxembourg: Publications Office of the European Union.

[9] Schwab, K. (2016). *The Fourth Industrial Revolution*. London: Penguin

[10] European Parliamentary Research Service (2015). "Industry 4.0: Digitalisation for productivity and growth". www.europarl.europa.eu/thi nktank/en/document.html?reference=EPRS_BRI%282015%29568337. Accessed 04.09.2020.

[11] Lundvall, B.-Å. (1996). "The Social Dimension of the Learning Economy". Aalborg, DRUID Working Paper No. 96-1.

[12] OECD (1994). "The OECD Jobs Study". Paris: OECD.

[13] Lundvall, B-Å., Rasmussen, P. and Lorenz, E. (2008). "Education in the learning economy: a European perspective". Policy Futures in Education, 6(2), 681-700.

[14] Bakhshi, H., Downing, J., Osborne, M. and Schneider, P. (2017). "The Future of Skills: Employment in 2030". London: Pearson and Nesta.

[15] Gregory, T., Salomons, A. and Zierahn, U. (2016). "Racing with or against the machine? Evidence from Europe". Technical Report 16-053.

[16] McKinsey Global Institute (2017). "A future that works: Automation, employment and productivity". Available at www.mckinsey.com/globa l-themes/digital-disruption/harnessing-automation-for-a-future-that-works. Accessed 04.09.2020.

[17] PwC (2016). "Industry 4.0: Building the digital enterprise". https://ww w.pwc.com/gx/en/industries/industries-4.0/landing-page/industry-4.0-building-your-digital-enterprise-april-2016.pdf. Accessed 04.09.2020.

[18] Deming, D. (2015). "The growing importance of social skills in the labor market". Technical Report 21473, NBER Working Paper.

[19] Lundvall, B-Å. (2014). "Deteriorating quality of work undermines Europe's innovation systems and the welfare of Europe's workers!" EUWIN Bulletin, June.

[20] Arundel A., Lorenz, E., Lundvall, B.-Å. and Valeyre, A. (2007). "How Europe's economies learn: a comparison of work organization and innovation mode for the EU-15". Industrial and Corporate Change, 16(6), 1175-210.

[21] Chesbrough, H. W. (2003). "Open Innovation: The New Imperative for Creating and Profiting from Technology". Boston: Harvard Business School Press.

[22] Desouza, K. C. and Awazu, Y. (2004). "Gaining a competitive edge from your customers: exploring the three dimensions of customer knowledge". KM Review, 7(3), 12-15.

[23] Prahalad, C. K. and Ramaswamy, V. (2004). "Co-creation experiences: the next practice in value creation". Journal of Interactive Marketing, 18(3), 5-14.

[24] Valkokari, K., Paasi, J. and Rantala, T. (2012). "Managing knowledge within networked innovation". Knowledge Management Research & Practice, 10, 27.

[25] Sudhoff, M., Prinz, C. and Kuhlenkötter, B. (2020). "A systematic analysis of learning dactories in Germany - concepts, production processes, didactics". Procedia Manufacturing, 45, 114-120.

[26] Totterdill, P., Pot, F. and Dhondt, S. (2016). "Definiowanie innowacji w miejscu pracy", In Strumińska-Kutra, M. and Rok, B. (eds.) Innowacje w miejscu pracy. Pomiędzy efektywnością a jakością życia społecznego, pp. 25-50, Warszawa: Poltext.

[27] Corrado, C. and Hulten, C. (2010). "How do you measure a technological revolution?" American Economic Review, 100(5), 99-104.

[28] Bolwijn, P. T., van Breukelen, Q. H., Brinkman, S. and Kumpe, T. (1986). "Flexible Manufacturing: Integrating Technological and Social Innovation". Amsterdam: Elsevier.

[29] European Commission (2014). "Advancing Manufacturing – Advancing Europe". Report of the Task Force on Advanced Manufacturing for Clean Production, SWD (2014) 120 final, Commission staff working document, Brussels.

[30] OECD (2010). "The OECD Innovation Strategy: Getting a Head Start on Tomorrow". Paris: OECD Publishing.

[31] Jensen, M. B., Johnson, B., Lorenz, E. and Lundvall, B-Å. (2007). "Forms of knowledge and modes of innovation". Research Policy, 36(5), 680-693.

[32] Tidd, J. and Bessant, J. (2009). "Managing Innovation: Integrating Technological, Market and Organizational Change". 4th edition, Chichester: John Wiley and Sons.

[33] Totterdill, P. (2015). "Closing the gap: 'The fifth element' and workplace innovation". European Journal of Workplace Innovation, 1, 1, January.

[34] Ennals, R., Holtskog, H., Berge, M., Midtbø, I. L. and Garmann Johnsen, H. C. (2018). "Coping with organisations: sociotechnical, dialogical and beyond". In Garmann Johnsen, H. C., Holtskog, H. and Ennals, R (eds.). "Coping with the Future. Strategies for Sustainable Development of Business and Work". London: Routledge.

[35] Pot, F. D. (2011). "Workplace innovation for better jobs and performance". International Journal of Productivity and Performance Management, 60(4), 404-415.

[36] Ramstad, E. (2009). "Promoting performance and the quality of working life simultaneously". International Journal of Productivity and Performance Management, 58(5), 423-436.

[37] Totterdill, P., Dhondt, S. and Milsome, S. (2002). "Partners at work? A report to Europe's policy makers and social partners". Nottingham: The Work Institute.

[38] Pot, F., Totterdill, P. and Dhondt, S. (2016). "Workplace innovation: European policy and theoretical foundation". World Review of Entrepreneurship, Management and Sustainable Development, 12(1), 13-32.

[39] Gustavsen, B. (1992). "Dialogue and Development". Van Gorcum: Assen/Maastricht.

[40] Lado, A. A. and Wilson, M. C. (1994). "Human resource systems and sustained competitive advantage: a competency-based perspective". Academy of Management Review, 19(4), 699-727.

[41] Huselid, M. A., Jackson, S. E. and Schuler, R. S. (1997). "Technical and strategic human resource management effectiveness as determinants of firm performance". Academy of Management Journal, 40(1), 171–188.

[42] Teague, P. (2005). "What is enterprise partnership?" Organization, 12, 567.

[43] Delery, J. E. and Doty, D. H. (1996). "Modes of theorizing in strategic human resource management: tests of universalistic, contingency and configurational performance predictions". Academy of Management Journal, 39 (4), 802-835.

[44] Sheehan, M. (2013). "Human resource management and performance: evidence from small and medium-sized firms". International Small Business Journal, published online 6 January.

[45] Business Decisions Ltd (2002). "New Forms of Work Organisation: The Obstacles to Wider Diffusion". Brussels: KE-47-02-115- EN-C, DG Employment and Social Affairs, European Commission.

[46] Appelbaum, E., Bailey, T., Berg, P. and Kalleberg, A. L. (2000). "Manufacturing Advantage: Why High-Performance Work Systems Pay Off". Ithaca, NY: ILR Press.

[47] ITPS (2001). "Enterprises in Transition: Learning Strategies for Increased Competitiveness".Östersund: ITPS.

[48] NUTEK (1996). "Towards Flexible Organisations". Stockholm: NUTEK.

[49] Oeij, P., Rus, D. and Pot, F. (eds.) (2017). "Workplace Innovation. Theory, Research and Practice". Cham, Switzerland: Springer.

[50] Exton, R. and Totterdill, P. (2019). "Unleashing workplace innovation in Scotland. International Journal of Technology Transfer and Commercialisation". 16(3). https://doi.org/10.1504/IJTTC.2019.099899

[51] OECD (2016). "The risk of automation for jobs in OECD countries: A comparative analysis". OECD Social, Employment and Migration Working Papers, No. 189. Paris: OECD Publishing.

[52] Oeij, P., Preenen, P. T., Van der Torre, W., Van der Meer, L. and Van den Eerenbeemt, J. (2019). "Technological choice and workplace innovation: towards efficient and humanised work". European Public & Social Innovation Review, 4(1), 15-26.

[53] Bakker, A. B. and Demerouti, E. (2007). "The job demands-resources model: state of the art". Journal of Managerial Psychology, 22, 309-328.

[54] Morgeson, F. P. and Humphrey, S. E. (2006). "The Work Design Questionnaire (Wdq): Developing and Validating a Comprehensive Measure for Assessing Job Design and the Nature of Work". Journal of Applied Psychology, 91, 1321-1339.

[55] Shantz, A., Alfes, K., Soane, E. and Truss C. (2013). "A theoretical and empirical extension of the job characteristics model". International Journal of Human Resource Management, 24(13), 2608-2627.

[56] Truss, C., Delbridge, R., Soane, E., Alfes, K. and Shantz, A. (eds.) (2013). "Employee Engagement in Theory and Practice". London: Routledge.

[57] CEDEFOP (2015). "Matching skills and jobs in Europe: Insights from Cedefop's European skills and jobs survey". http://www.cedefop.eu ropa.eu/en/publications-and-resources/publications/8088. Accessed 04.09.2020.

[58] De Sitter, L.U. (1981). "Op weg naar nieuwe fabrieken en kantoren" (Translated: Heading for New Factories and Offices). Deventer: Kluwer.

[59] De Sitter, L. U., den Hertog, J. F. and Dankbaar, B. (1997). "From complex organizations with simple jobs to simple organizations with complex jobs". Human Relations, 50(5), 497-534.

[60] Stieber, T. (2015). "I, Cobot: Future collaboration of man and machine". The Manufacturer (2015-11-15). https://www.themanufacturer.com/articles/i-cobot-future-collaboration-of-man-and-machine/ Accessed 04.09.2020.

[61] West, M. (2012). "Effective Teamwork: Practical Lessons from Organizational Research". Oxford: Blackwell.

[62] European Foundation for the Improvement of Living and Working Conditions (1997). "Employee Participation and Organisational Change. EPOC survey of 6000 workplaces in Europe". Dublin: European Foundation.

[63] West, M. A. and Lyubovnikova, J. (2012). "Real teams or pseudoteams? The changing landscape needs a better map". Industrial and Organizational Psychology, 5, 25-28.

[64] Mumford, E. (2006). "The story of socio-technical design: reflections on its successes, failures and potential". Information Systems Journal, 16, 317-342.

[65] Ghiselli, E. and Siegel, J. P. (1972). "Leadership and managerial success in tall and flat organization structures". Personnel Psychology, 25(4), 617.

[66] Boud, D., Cressey, P. and Docherty, P. (2006). "Productive Reflection at Work". London: Routledge.

[67] Buchanan, D. A. and Dawson, P. (2007). "Discourse and audience: organizational change as a multi-storey process". Journal of Management Studies, 44(5), 669-686.

[68] Høyrup, S., Bonnafous-Boucher, M., Hasse, C., Lotz, M. and Møller, K. (eds.) (2012). "Employee-Driven Innovation: A New Approach". London: Palgrave Macmillan.

[69] Senge, P. (1990). "The Fifth Discipline: The Art and Practice of the Learning Organization". New York: Doubleday.

[70] Caldwell, R. (2005). "Agency and Change: Rethinking Change Agency in Organizations". Abingdon: Routledge.

[71] Gronn, P. (2002). "Distributed leadership as a unit of analysis". Leadership Quarterly, 13,4, 423-451.

[72] Appelbaum, E. and Batt, R. (1994). "The New American Workplace: Transforming Work Systems in the United States". US: ILR Press.

[73] Gregory, D., Huzzard, T. and Scott, R. (2005). "Strategic Unionism and Partnership: Boxing or Dancing?" Houndmills: Palgrave MacMillan.

[74] Cressey, P., Exton, R. and Totterdill, P. (2013). "Workplace social dialogue as a form of 'productive reflection'". International Journal of Action Research, 9(2), 209-245.

11

Competencies in Digital Work

Matti Vartiainen

Department of Industrial Engineering and Management, Aalto University, Espoo, Finland

Abstract

Decisions to integrate digital technologies into work processes have a wide influence in work organizations and needed competencies, now and in the future. This chapter concentrates on exploring how digitalization is related to present and future competencies by transforming work processes, task and job contents, and the organization of labor, and finally products and services. Changes in work processes produce changes in job and task structures: jobs and tasks are replaced and destroyed when human labor is removed, hybridized when new features and demands are added, and recreated when new, previously unseen work requirements emerge. This, in turn, creates the need for reorganizing. In this way, digitalization increases pressures to rearrange work system elements anew. In addition to new ways of working and leading, new competencies are needed. Future-oriented competencies are needed at the individual, organizational, and societal levels. They include competencies to adapt and create new ways of working, anticipation, and digital competencies.

Keywords: Competency, digitalization, work process, job demands, renewed job and task contents, resilience

11.1 Introduction

11.1.1 What Happens to Work, Now and in the Future

Converting information to a digital format, digitalization of tools, products and services, value-adding processes, working environments, and the adoption of digital business models gradually change the nature of work and ways of working in micro-, small-, medium-sized, and large companies, in local regions such as cities, and globally on virtual online platforms. This development is shown as changes in the work settings or work environment, and where and when work is done. For example, employees increasingly act remotely, are mobile, and work and collaborate from multiple locations, virtually [1, 2, 3] enabled by digital collaboration technologies, as during the COVID-19 pandemic. *This chapter concentrates on exploring how digitalization is related to present and future competencies[1] by transforming work processes, objects of work, task and job contents, and the organization of labor.* However, in addition to technologies, there are many other factors that influence competencies needed in the future, such as climate change requiring new behaviors, the increasing heterogeneity of societies, and the need for resiliency in work and life because of unexpected changes in environments and individual value preferences. To illustrate possible future trajectories, examples of this development are given especially from the viewpoint of potentially needed competencies. The approach is exploratory and descriptive, as the evidence is mainly acquired from the available secondary material such as the literature and white papers in addition to some empirical studies.

During the past decade, the computerization of occupations in offices and industry [4] has dominated discussions on the future of work with claims of task and job replacements or their transformation into new types of work. Simultaneously, the sharing economy, or global platform economy, has become a prominent topic in working life rhetoric presenting crowd work [5, 6] and online gig work [7] on global online platforms as new types of jobs. Howcroft and Bergvall-Kåreborn [5] describe crowd work functions as a marketplace for the mediation of both physical and digital services and tasks. For digital tasks, the entire activity is carried out online, from initial instructions to completion and evaluation. Physical tasks are managed and mediated digitally (often via an app) but carried out offline (e.g., transportation of food). The computerization of work has become apparent in the social interactions of individuals (i.e., on the microlevel), in groups and communities

---

[1]"Competence" refers to those characteristics, knowledge, and behaviors a person has and uses, whereas a "competency" refers to those (s)he needs. Both concepts are used in this chapter; one to refer "competence in use," another to "competency needed." See definitions in Chapter 4.

(i.e., on the mesolevel), and in societies (i.e., on the macrolevel). On the microlevel, smart technologies—such as wearables using sensors to measure the quality of sleep and health—intrude on the life of an individual or a small group of individuals in a particular social context. On the mesolevel, which falls between the micro and macro levels (such as a community, organization, or city), technology takes the form of smart transport systems, service applications, social media, and various applications of the Internet of Things (IoT). At the macrolevel, technology (such as global work platforms) potentially influences the outcomes of interactions (such as economic transactions or other resource transfer interactions) over a large population. Flexible ways of organizing work have become more common along with the delocalization of work on a global scale.

Lehdonvirta [8] suggests that there are at least two types of delocalization: work that is partly relocalized elsewhere and delocalized work that is dispersed. Intraorganizational examples of *dispersed, delocalized* work units are now traditional distributed virtual teams (VTs) and projects. From the organizational viewpoint, they were the first birds migrating to the working life. VTs have been widely studied since the late 1990s (e.g., [9]). In these kinds of organizations, dispersed employees are still formally attached to their employer. Other forms of *dispersed* work are based on the idea of crowdsourcing labor input, which is obtaining needed services, ideas, and content by soliciting contributions from a large group of detached people—and especially from an online community—rather than from traditional employees or suppliers. In this way, small and big companies integrate (at least temporarily) self-employed workers ("freelancers") as independent contractors into their value chains. Virtual or digital work indicates, e.g., working as a freelancer on the Internet doing microtasks from a remote place, utilizing digitalized working platforms. As described by Lehdonvirta [8], this type of digital dispersed work is organized by platform, such as Amazon's Mechanical Turk. An employer posts small digital tasks for the site's users to complete. A worker enters the site using an owned or borrowed device, selects a task, completes it, gets credited with the proceeds, and selects the next task. Each completed task usually earns the worker-user a small remuneration. Because the work is detached from local institutions, in this chapter I call these workers a special type of tele- or remote worker, that is, "detached temporary global teleworkers."

To meet present and future challenges, new competencies are needed, and old ones must be updated for the resilience and "sisu" of working individuals, organizations (employers), and society at large. Resilience connotes organizational, team, and individual capacities to absorb external shocks and to learn from them, while simultaneously preparing for and responding to external jolts. The Finnish concept of "sisu" [10] refers to the enigmatic power that enables individuals to push through unbearable challenges. Resilience can appear in two differentiated ways, complementing each other as an adaptive or a reactive response to external jolts and stressors [11]. With organizational environments becoming more unstable, uncertain, and equivocal in recent years, the concepts of resilience and "sisu" are increasingly significant for public administration, small and big organizations in various working life sectors, and their employees. The remainder of this chapter is organized in the following manner. First, the role of emerging technologies as a driving force of organizational development is defined. Secondly, mechanisms are described through which decisions to digitalize influence work processes and increase the pressure to renew work systems. Digitalization penetrates deeply into the work environment, changes task and job contents, and creates the need to reorganize work and work relations. Thirdly, concepts of competency/competence are defined. Finally, some implications concerning the future competencies of digital labor are presented.

11.2 Technology as a Driver for a Change in Work

Of the general global trends, the most immediate material factor affecting people's work activities and resources on the organizational level is the extensive utilization of digitalization, including the mobile Internet. Key applications in communications, working platforms, and the automation of work processes affect work and leisure and their relationship in many ways. Digital platforms and the transition to online virtual work, the analysis and algorithmization of large bulk data into intelligent cloud services, "artificial intelligence," the "Internet of Things," and the "mobile Internet," machine learning, and robotization are particularly associated with changes in the structure and contents of work—and possibly competencies. In addition to changes in job content and structures, many anticipate that tasks and jobs will completely disappear (e.g., [4]), although critics consider this claim exaggerated (e.g., [12]).

11.2.1 Waves of Technological Development

The role of technology in working life has strongly increased since the "computerization" of work by means of computer-controlled equipment, which began in the 1970s. However, its influence has pervaded throughout as technological innovations play a crucial role in the development of economies and whole societies. Technologies not only have a "push" impact on everyday life, but they are strongly "pulled" by economic interests and decisions to take technologies into use. The long-term growth of the economy, including crucial changes in the production and organization of work, has been theorized to occur as "Kondratieff waves'" [13] or "great surges" [14]. The "long wave theory" (e.g., [14]) argues that there are repeating phases over 50–60 years that are associated with technological innovations. Kurki and Wilenius discuss this topic in more detail in Chapter 2 in this book.

According to this theory, the present "computerization" or digitalization is just one phase in the long developmental chain. Kondratieff [13] in the 1930s and several other scholars found that the British industrial growth accelerated around 1780, primarily in the cotton and iron industries. Driven by water power, this rapid growth constituted the first great surge of the "industrial revolution," which lasted from approximately 1780 to 1840. The second surge revolved around the steam engine and railways, while the third revolution is known as the age of steel, electricity, and heavy engineering. The fourth surge took place in the age of oil, the automobile, and mass production. Finally, the fifth surge, the "Age of Information and Communications," occurred in the 1970s with chips, hardware, and new materials, followed by software and telecommunications equipment, laptops, e-services such as e-mail, the Internet, and new forms of electronic communication. The fifth wave was also driven by digitalization: the personal computer, automation, and improved communication. The present surge—called the sixth wave of "Intelligent Technologies" [15: 9]—started around 2008: "During the next wave, our economies will be driven by environmental technologies, biotechnology, nanotechnology, and health care. Their effect is leveraged by digitalization and the exponential rise of computational power—both legacies of the previous wave—that create circumstances for new products and services."

11.2.2 Types of "New" Technologies

It has been said that there is no "new" technology, as technologies develop incrementally, although there are radical technological inventions that leverage future technologies [16]. For example, Linturi and Kuusi [17] in Chapter 3 of this book (with web links) describe 2000 recent technological breakthroughs. Many of them are used for a very specific purpose, such as increasing safety with a radar for bicycles. In any case, the ability to produce, store, process, and transmit digitally coded information has grown exponentially in the last few decades. The microprocessor is still considered the key technology behind digitalization (e.g., [18]). This empirical observation has been justified by referring to Moore's law [19]. In 1965, Moore predicted a doubling of transistor density every year for at least the next 10 years. In a 2010 interview[2], Moore himself admitted the limits of his law "In terms of size [of transistor], you can see that we're approaching the size of atoms, which is a fundamental barrier." However, it is fascinating to read about his visions [19: 114].

> *"Integrated circuits will lead to such wonders as home computers or at least terminals connected to a central computer, automatic controls for automobiles, and personal portable communications equipment. The electronic wristwatch needs only a display to be feasible today. But the biggest potential lies in the production of large systems. In telephone communications, integrated circuits in digital filters will separate channels on multiplex equipment. Integrated circuits will also switch telephone circuits and perform data processing. Computers will be more powerful, and will be organized in completely different ways. For example, memories built of integrated electronics may be distributed throughout the machine instead of being concentrated in a central unit. In addition, the improved reliability made possible by integrated circuits will allow the construction of larger processing units. Machines similar to those in existence today will be built at lower costs and with faster turn-around."*

In 1975, looking ahead to the next 10 years, he updated his estimate to a doubling every 24 months [20]. In all, the capacity and power of various technologies have been swift ever since. The "fifth wave" brought computers to the home and mobile phones to the pocket, and the "sixth wave" has made them connected and smart, penetrating both working life and free time. It has been claimed that digital technologies change economies by the automation of work, digitalization of processes, and coordination by platforms [18].

Automation of work by replacing human work in production and distribution processes is not a new phenomenon. Figure 11.1 nicely

---

[2]https://www.techworld.com/news/tech-innovation/moores-law-is-dead-says-gordon-moore-3576581/

Figure 11.1 The spectrum of automation [23].

illustrates the principle of how technological innovations spread by blending and merging microlevel applications into the greater wholes. The development seems evolutionary rather than revolutionary. For example, robotics is not a new phenomenon, but robots todays are advanced, being much more flexible than before because of better senses (sensors) and much more intelligence (software algorithms, processing capacity) [17]. In manufacturing, technologies such as three-dimensional (3D) printing or additive manufacturing promise to turn the world of producing physical objects into fully personalized on-demand manufacturing [21]. This development could bring back home off-shored work from faraway countries—and at the same time, it influences structures of work processes and organizing. With the Internet of Things, there emerges an increasingly complete virtual copy of our physical world, which in turn enables new possibilities to collaborate [22].

Digitalization of processes is shown in the development of sensors, which are used to translate the physical production process into digital information. In this translation, work phases are removed and replaced by new increasingly smart hybrid digitalized phases and operations. Sensors have become cheap, readily available, and more powerful than before. As a taste of what may be on the horizon, sensor technology has already been employed in cases such as Japan's nuclear disaster, where hackers devised a grid of self-made Geiger counters that were connected to an online map. In Chile, sensors give early warning alerts about possible earthquakes. Sensors are widely used to collect "Big data," which is analyzed to build digital services, such as offering health advice. A vast amount of information is gained from data transfers on the Internet and end users' smart devices and their behavior.

The Internet of Things is one of the key technologies as it enables the connection of all the "things" having an IP address (such as refrigerators, clothes, and cars) to each other via the Internet. Information from those devices will then be a central part of the basic infrastructure and will blend in with the physical reality. At the same time, the technology has reached a level where it is technically available to large groups of people. This combination is the brew in which experimentation and innovation take place.

Online platforms are digital spaces that often exploit cloud technologies and big data and its analytics. The concept of "platform" is related to "platform economy," which refers to the mediation between the supply and demand for work, goods, and services through an online platform. Riso [24] describes the plethora of close concepts such as the "sharing economy," "collaborative economy," "gig economy," and "crowdwork," to name a few. Online platforms enable fully digital online work and coordination of economic transactions. Cloud computing enables the storage of data in networked data centers, and its processing and distribution as applications and services for individuals and organizations [25]. For example, the cloud includes webmail services and, for instance, collaborative text editing in Google Drive and collaboration in MS Teams. The cloud is enabled by the increased speed of broadband connections. Before, accessing large amounts of data remotely would have been too time consuming, thus crippling the process. One key element of the platforms is a set of algorithms for matching and coordinating transactions automatically. The algorithms provide a governance structure to the platforms, incorporating encoded rules as well as automated monitoring and enforcement mechanisms. In practice, numerous platforms for different purposes are available. They can be privately owned (e.g., Airbnb) or owned by their users (e.g., blockchain). Platforms can be used for the exchange of goods (e.g., Amazon Marketplace) or services such as Uber.

It can be concluded that technological development—with thousands of incremental and some radical innovations—potentially impacts value processes and products and services as their outcomes. However, the decisions to develop, implement, adapt, and use are made by human actors. It is important to notice that the development is not deterministic but is ultimately based on human needs and values.

11.3 Digitalization Has a "Long Tail"

Digitalization penetrates the components of sociotechnical work systems and their contexts to varying degrees, generating the need to update human practices and competencies. Figure 11.2 shows the logic of this penetration as a "snowball effect": a technology affects a work process by transforming the microstructure of its phases, which is then mirrored in other components

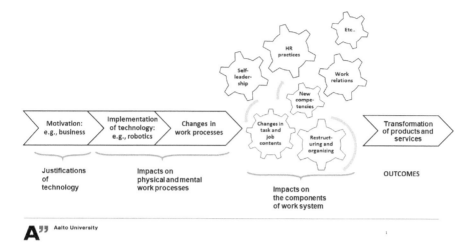

Figure 11.2 The logic of digitalization influence on work systems.

of the work system. The phases and operations of a value creation process are either removed, hybridized, or renewed, driven by decisions to implement and use digital smart technologies. As the value-producing process is usually still steered and managed by human activities, the changes are also needed in human tasks and jobs. Through this process of transformation, digitalization also replaces, hybridizes, and renews tasks and job contents. This, on its behalf, leads to a need to reorganize and structure organizations. As the consequence, many other organizational and human-related issues are under pressure and need for changes. For example, a restructuring and organizing of work units are needed, resulting in new ways of working, leadership practices, and competencies. The outcomes of this transformation are digitally influenced products and services. This chapter shows the logic in more detail.

11.3.1 Transforming Work Processes

The implementation of technologies affects work processes directly. It is predictable that computerization and digitalization during the next decades will change many value-creating processes. In manufacturing, the implementation of robots and 3D printing illustrates this sociotechnical change. Robots can assemble products, whereas 3D printing can bring spare parts manufacturing back to local production sites. 3D printing is a process where a physical object is created from a computer model through

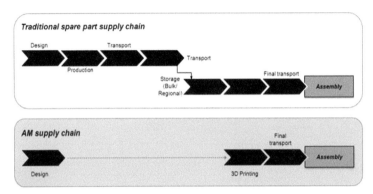

Source: The new software-defined supply chain. IBM Institute for Business Value. 2013, 15 p.

Figure 11.3   The new software-defined supply chain: the influence of using 3D printing in spare part production [26].

successive material layering. Producing spare parts has traditionally required organizations to invest a lot of time and money in estimating the required amount of spare parts, producing these parts, storing, and finally transporting them to the customer, often from faraway places. Utilizing 3D printing as an alternative method for manufacturing is attractive because of the flexibility, lower costs, and speed that it offers.

For example, the development of 3D printing and its use in spare parts production can remarkably reduce the throughput time (Fig. 11.3) by removing former time-consuming phases from the work process. At the same time, however, phase-related human operations and actions are removed. For example, the transport of spare parts by truck, driven by a skilled driver, is no longer needed. At the same time, it will be possible to bring back manufacturing nearer to the assembly, e.g., from abroad.

Another example of human task replacement is "Gordon," a robot barista[3] that makes a caffé latte in a minute, with a price less than in an ordinary café (Fig. 11.4). The latte is ordered and paid remotely via smartphone

---

[3]https://www.handelskraft.com/2017/02/premium-coffee-cups-served-by-robots-in-seconds-5-reading-tips/

Figure 11.4 Gordon—a robot barista.

using an app or at the coffee shop via laptop. The robot Gordon receives the order, prepares a coffee in less than a minute, and when it is ready, it sends an SMS with a four-digit code to the customer waiting at the robot's coffee shop (or kiosk). Then, the robotic arm grabs a cup of coffee from one of the eight warming stations and gently places it on a small shelf where it is picked up. Gordon is able to prepare 100–120 cups an hour. The remaining human tasks with Gordon are cleaning, chopping ingredients, and maintenance. Service robots such as Gordon could have much influence in the future on employment in the restaurant business. The National Restaurant Association (2017) from the United States estimates that nearly 1 in 10 American workers are employed in the restaurant business, numbering 14.7 million employees. However, the progress of change maybe slower, as people still enjoy chatting with a human barista while waiting and drinking their coffee.

Changes in work processes can influence the contents of tasks and jobs. In their seminal and dystopic work, Frey and Osborne ([4], see also [27]) estimated the probability of computerization for 702 occupations in the United States. They focused on technological advancements in machine learning (ML), including data mining, machine vision, computational statistics, and other subfields of artificial intelligence (AI). Especially in AI, algorithms are developed that allow cognitive tasks to be automated. In addition, they examined the application of ML technologies in mobile robotics (MR) and computerization in manual tasks. According to their estimates, around 47% of employment in the United States over the next decade or two is in the high-risk category, replacing most workers in the transport and logistics occupations, together with office and administrative support workers, and labor in production occupations. They also found that a substantial share of employment in service occupations is highly susceptible to computerization because of the increasing number of service robots. They conclude that as technology races ahead, low-skill workers will be reallocated to tasks that are not susceptible to computerization, i.e., tasks requiring creative and social intelligence. For workers to win the race, however, they will have to acquire creative and social skills and competencies. Pajarinen et al. [22] repeated the study using the same methodology and also analyzed data from Finland and Norway. Their replication of Frey and Osborne, using data for 2012 rather than 2010, suggests that 49% of the US employment is in the high-risk category. The corresponding share for Finland is 35% and for Norway 33% (i.e., 14–16% less than in the United States). Low-wage and low-skill occupations appear to be more threatened, whereas service and public sector jobs are relatively more sheltered than those in the manufacturing and private sectors.

However, critics—cited by Rani and Grimshaw [28]—argue that not all occupations are at high risk or will be automated and that only a handful of tasks within particular occupations will be automated. This focus on tasks finds a lower risk of job losses, varying from 6 to 15%, depending on the country. Also, Méda [12] remarks that the negative effects of digitization on employment should not be taken for granted just yet and should be considered as technologically deterministic. She refers to empirical studies showing that increased robotization did not affect employment and even created new jobs, and the workforce involved in the "gig economy" is still small, even though increasing continuously. In addition, what holds for a particular industry segment does not apply to all sectors. These studies also forget that substituting humans with robots is not the only solution. An equally plausible option would be that of human–robot collaboration, or "cobotization," [12] offers a symbiosis between work done by humans and robots.

In all, it can be concluded that different technologies, digital online platforms as workplaces, and digitalization of work generally based on human decisions have in different sectors specific impacts on task and job content and work structures, and through them on competencies. The decisions to implement and use digital innovations can:

- *Replace and destroy tasks and jobs* by removing human labor in work processes. For example, robotics and 3D printing replace some phases and human operations in manufacturing processes, or algorithms replace some cognitive functions? Old competences are not needed as tasks disappear, the expected societal outcomes are unemployment or an "old" workforce with new competencies working in another job.
- *Hybridize tasks and jobs* by adding new features and demands to them, e.g., medical diagnosis with the help of AI, leading to enriched and hybrid jobs, "cobotization"? partly new competencies, new constituents in competencies, renewed job descriptions, e.g., digital marketer, re-education of employees for keeping the jobs.
- *Create new tasks and jobs* by adding completely new job demands and reallocating jobs, e.g., work in social media, advisors in virtual worlds? Completely new competencies, new job descriptions, e.g., platform engineer, new jobs are created, a workforce is needed.

11.3.2 Changing Organizational Forms and Structures

The impacts of digitalization do not stop here, as they reflect on how to organize and manage work relations. The remaining hybrid and newly created tasks and jobs need restructuring and reorganizing. Some "old" new ways of working (such as virtual teams) can be used as benchmarks for present and future new ways of digital working, as the use of information and communication technologies characterizes them too. Studies on virtual teams and projects have revealed five characteristics that also describe current upcoming forms of organizing: locational multiplicity, structural flexibility, temporal flexibility, time used in digital communication, and diversity of workers.

Work is more often *done in multiple places* outside the main workplace, supported by mobile information technology. The COVID-19 pandemic showed the massive shift of knowledge workers to remote work. The *flexible structures* of working life involve organizing assignments into temporary projects for which a group of workers is assembled as necessary to realize the project at hand. Virtual dispersed teams are used as a standard way of cooperating. The main trend is to build flexible, adaptable forms of organizations. More often, work is organized within and between workplaces into *temporary projects*. Also, individual assignments are often temporary, especially when working on platforms. In addition to the variation in places, flexible ways of working also involve temporal flexibility, that is, a worker can or is expected to begin and end his or her work according to the situation and the need. In addition to temporal flexibility, the time spent in working, communicating, and *collaborating with digital tools* is increasing. Colbert et al. [29] refer to studies on "digital natives" that show that within five minutes of waking up, at least 25% of teenagers have reached for a smartphone or other electronic device. Children between 8 and 12 use screen media an average of slightly more than 4.5 hours each day, while teens (ages 13 through 18) average slightly over 6.5 hours on screens. According to a study conducted by Nokia, the average American smartphone user checks their phone every 6.5 minutes, or up to 150 times per day [30].

From the competence viewpoint, studying digital natives and "heavy adult users" in ways of using and practicing a trade with these tools might be one way to forecast future competencies. Additionally, technology blurs the lines between work and nonwork domains. Through the Internet, including social media, it is easy and usual to have access to family, friends, and shopping. On the other hand, technology makes it possible for employees to remain connected to work from when they are at home [3].

11.3.3 Creating New Ways of Digital Working

New ways of working have been characterized as changes to how work is done, where it is done, and by whom [31]. These ways are also examined from the perspectives of an individual, a team or a project, an organization, and as a societal phenomenon. These flexible and temporary modes as emerging forms of employment [2], often initiated by technologies, are also characterized by unconventional work patterns and places of work, or by the irregular provision of work.

11.3.3.1 Digitally Enhanced Remote Work

When work assignments are done outside the "main place of work," it is referred to as telework or telework. Thus, telework is defined as "working outside the conventional office using telecommunication-related technologies to interact with supervisors, co-workers and clients" [32]. When employees are changing their working locations daily or weekly, they are doing mobile and multilocational work (e.g., [1, 33]). Fully digital work has appeared detached from any stabilized social and organizational settings as *detached temporary global telework*. The development of digital working environments, such as global online platforms employing microproviders [34], has expanded working locations worldwide. Detached global teleworkers often work from their homes on different continents. Remote workers often collaborate from their present locations in temporary or permanent virtual teams and communities. Mobile Internet and cloud services enable mobile, multilocational work. Cooperation with others is done in local meetings, online, and periodically at the main workplace, if it exists. Additionally, when work is done with others in virtual or digital collaboration, it is usually traditionally referred to as virtual teamwork.

11.3.3.2 From Mobile, Multilocational Work to Global Virtual Collaboration

The second main type of the now traditional "new ways of working" is virtual collaboration in distributed teams and projects. Distributed or dispersed work also has deep historical roots [35]. Team members working in different locations and their geographical distances from each other constitute a distributed team [36]. These teams become virtual when extensively utilizing various forms of computer-mediated communication that enable geographically dispersed members to coordinate their individual efforts and inputs. Virtual teams have three main attributes [37]: (1) A functioning team with interdependent task management, having shared responsibility for outcomes, and collectively managing relationships across organizational boundaries; (2) team members are geographically dispersed; and (3) they rely on technology-mediated communication. Social space partly overlaps virtual space because people communicate and socialize through virtual interaction.

Physical mobility and multilocality of team members add a new set of requirements to distributed work by providing the varying space aggregations as working contexts. Virtual teams with mobile, multilocational members are always distributed. But not all distributed, virtual teams are mobile, because their members may work from fixed locations using virtual spaces. The use of virtual spaces for communication and collaboration makes a team into a distributed virtual team or mobile virtual team. Thus, the main types of nonconventional teams are (1) distributed, (2) virtual, and (3) mobile virtual teams. They differ from conventional face-to-face groups and teams specifically in three characteristics: an increased *use of different dispersed physical spaces* by their members, the use of *virtual space* also for social interaction, and *physical mobility,* which brings continuously changing aggregations of spaces as working contexts.

11.3.3.3 Working on Digital Online Platforms

Digital online platforms serve more and more as crowdsourced dispersed workplaces. A "platform" is a "palette" consisting of usable components to be used for different purposes by one or several actors. Platforms as the working context can be global and/or local. Any work platforms (e.g., Amazon Mechanical Turk) act as employment agencies. An employer somewhere in the world digitally posts tasks for the site's users to complete. A worker enters the platform using their own or a borrowed device, selects a task, completes it, gets credited with the proceeds, and selects the next task. Each completed task earns the worker remuneration.

An entire "digital working class" has emerged worldwide, working for international employers [38]. As described earlier, digital delocalized labor is characterized as two types: work as partly relocalized elsewhere and delocalized work that is dispersed [8]. *Relocalization* is exemplified by offshored work and *dispersed* work by globally crowdsourced microwork. Virtual or digital work indicates working as a freelancer on the Internet doing microtasks from a remote place utilizing digitalized working environments. A major difference from conventional jobs is the fact that this kind of work is entrepreneurial freelancing, often without any kind of social protection. Lehdonvirta [8], who has studied microwork and their platforms, states that the key feature of these platforms is that they provide employers with an application programming interface: a codified interface through which the employer's software can issue inputs to and receive outputs from the workforce as if it were a software module. Employees doing microwork maybe detached from interactions with other employees and from national legal frameworks and contracts.

Digital technologies replace a part of the tasks traditionally performed by people, but they also create new, partly more demanding knowledge work. The most familiar digital services are bank services, which have already become online services for the majority of the population.

Digital online platforms and crowdsourcing platforms as new working environments are varied, from those based on volunteer work (e.g., Wikipedia) and those exploiting it (e.g., Facebook, Google, YouTube) to global platforms providing employment service (e.g., Amazon Mechanical Turk) and enterprises using internal social media. Routine microtasks (e.g., translating an advertising slogan from Finnish to English), as well as large innovation projects, are carried out on the digital working platforms. Perhaps the best example of a knowledge community exploiting voluntary work is Wikipedia, which provides a free-of-charge information pool for all with access to the Internet. For example, Facebook, Google, and YouTube, in turn, exploit voluntary peer production effectively to promote their own business. An example of a Finnish working platform based on common interests is Solved (https://solved.fi/), which provides a working environment to 4,000 cleantech experts and consultation services utilized in new and experimental design projects.

11.3.4 Loose Employment Relationships

As a consequence of the transformation of work tasks and jobs and reorganizing, the supply and demand of the workforce will be polarized— there is an oversupply of some skilled workers and an undersupply of others. Luckily also now jobs with new types of competencies are created. Unpredictability will, however, increase both in organizations and on the personal level. The ways of working will change within and between organizations. Organizations will adapt their operations to the demands of the market and to seasonal variations by means of flexible employment relationships.

The types of employment relationships from the perspectives of an individual worker's working life cycle and an organization are illustrated in Fig. 11.5 as the core and contingent types of the workforce, such as temporary external workers and freelancers whose labor is purchased.

Contingent and autonomous types of jobs are sometimes called "flexible" (e.g., [2]). Increased flexibility demands have resulted in the emergence of new forms of employment across the world. These have transformed the traditional one-to-one relationship between an employer and an employee. Kellogg et al. [39] describe the potential role of algorithms to reshape the relationship between employers and workers leading to "algorithmic management" [40]. They claim that employers look for economic value and benefits of algorithmic technologies based on improved efficiency in decision-making, coordination processes, and organizational learning. They argue that employers can use algorithms to control workers through six main mechanisms, which they call the "6 Rs": employers can use algorithms to help direct workers by restricting and recommending, evaluate workers by recording and rating, and discipline workers by replacing and rewarding. However, employees have several tactics to resist unwanted changes in work.

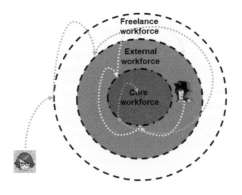

Figure 11.5 Types of employment relationships from an individual worker's and an organization's perspectives.

They can individually resist by not cooperating, by leveraging algorithms, and by personally negotiating directly with clients. They can also resist through collective actions in online forums by rating employers who treat their employees badly, as well as building platform cooperatives. New professional codes of ethics have also been built in addition to legal renewals in different countries, especially concerning platform workers as independent contractors or employees.

From an organization's perspective, the *core workforce* consists of workers indispensable to the organization and in a permanent employment relationship, with the expertise, know-how, and professional competencies. The core workforce is supplemented by a *contingent workforce* in a looser relationship with the organization, performing routine wage work of lesser value to the work organization. The contingent workforce, more often than the core workforce, is in the so-called atypical employment relationships, characterized by part-time employment and zero-hour contracts, for example. The *freelance workforce* is characterized by an occasional, result-based relationship with the contracting organization. The workers are not the organization's own workforce; they are (e.g.,) employed by a subcontractor or a work lease enterprise. They also include freelancers, consultants, and contract workers. The fixed-term nature applies to almost all occupations. Many tasks are seasonal (peak periods and seasons, night and weekend work, substitutions, etc.). According to Eurostat, 14% of the employed workforce (including the agricultural sector) was self-employed in 2017.

Lehdonvirta [8, 41] from the Oxford Internet Institute states that it is crucial for the crowdsourced business model to have labor detached from local organizational and institutional structures, so that it maybe sold as an on-demand "cloud" service. In his study on microwork [8], he found that microwork is seldom "a full-time job." For example, the Nepalese workers mainly depended on their parents for their subsistence and used microwork to earn additional income. The Filipino workers were "precariats," combining microwork and other irregular income sources. The American workers were a mix of precariats (typically housewives economically dependent on their spouses) and salary earners who did microwork as a hobby as much as to earn additional income. He also reports about the location of the online workforce [41]. According to the institute's data, the largest overall supplier of online labor is India, which is home to 24% of the workers observed. India is followed by Bangladesh (16%) and the United States (12%). Different countries' workers focus on different occupations. The software development and technology category is dominated by workers from the Indian subcontinent, who command a 55% market share. The professional services category, which consists of services such as accounting, legal services, and business consulting, is led by UK-based workers with a 22% market share.

11.4 Competencies in Digital Work

Digitalization influences work processes, task and job contents, and how they are organized, in addition to job demands in the work environment. The change is systemic, indicating dependencies on the interaction and dynamics of the system components. These changes further reflect on individual-, team-, and organization-level competencies. They are needed as intangible resources to meet changing job demands and to regulate needed activities, actions, and operations in the interaction with the environment to produce products and services. To identify competencies, the changed and changing processes in their contexts need critical examinations of the competencies' validity. In the case of future competencies, their contents are anticipated using scenarios of the future demand for jobs. To clarify the conceptual difference, the "competency" and the "competence" and their manifestations on different organizational levels need operationalizations. They enable identifying competencies needed in the digital future.

Table 11.1   Some basic definitions of competenc(i)es.

Spencer & Spencer [44, 29]	"Competencies have been defined as learned capabilities that lead to effective or superior performance and are reflected by a set of behaviors that share a common underlying intent" [45]
Bertram & Roe [46]	"A learned ability to adequately perform a task, duty or role" [55, 195]. Competences typically integrate knowledge, skills, personal values, and attitudes; and they build on knowledge and skills and are acquired through work experience and learning by doing.

11.4.1 What are "Competency" and "Competence"?

Differences between *competencies needed* in work and *competences in use* show that the digitalization is an evolving phenomenon. Both competences in use (i.e., competence) and competencies needed (i.e., competency) refer to intangible resources used and needed when working in present and future digital work systems and environments. Discussions on "competencies" and "competences" have long continued; however, there still remain some conceptual ambiguity, a lack of methodological rigor, and a dubious psychometric quality in their measurement, as shown by Stevens [42]. For example, Teodorescu [43: 30] notes, "there are many differences and some similarities between competence models and competency models in their definitions, areas of focus, and applications. The bottom line, though, is that organizations pay people for results, not behaviors, and to ensure that your organization is training, supporting, and developing the right competencies, skills, knowledge, and behaviors, you must first define what competence is for each role." Some basic definitions of competenc(i)es are shown in Table 11.1.

To conclude, the term "competency" refers to those skills, knowledge, and behaviors that are *needed* in a job, describing how things could be done and at what level, whereas the term "competence" refers to a person's *actual* knowledge, skills, and behaviors that he or she possesses and *potentially uses* when acting. The terms differ in their work and worker orientations. This difference matters when collecting and analyzing empirical data in research. In this chapter, the term "competency" is used as an integrated view of competencies and competences consisting of two main elements: the underlying characteristics or attributes and the operative actions or behaviors that are needed in work. For example, in interviews, interviewees are asked about "required" and "needed" competencies in a digital work setting.

The level of needed competencies also varies. The needed competencies are divided into the micro-, meso-, and macrolevels:

- *Individual competency* denotes a needed characteristic of a person that results in effective performance on the job [45:21, 44]. Competencies are learned abilities to perform a task, duty, or role in a particular work setting, integrating several types of knowledge, skills, and attitudes [48, 47].
- *Team competency* refers to those collective characteristics and processes that a team must have to operate in a certain work environment. A collective competence is a group's ability to work together toward a common goal. This includes the group's ability to solve problems together and interpersonal competencies to work together with different individuals [49, 50], and knowledge and a repertoire of procedures shared by a team or a project in their work context.
- *Strategic core competency* reflects resources and capabilities that a whole organization must achieve to maintain a competitive advantage [51, 52, 53] on the macrolevel.
- *Societal competencies* refer to generic competencies needed in a society and also globally. These competences are needed for the sustainable development of societies, e.g., digital competences [54].

Athey and Orth [55] synthesize a competency widely to include "... a set of observable performance dimensions, including individual knowledge, skills, attitudes, and behaviors, and collective team, process, and organizational capabilities, that are linked to high performance and provide the organization with [a] sustainable competitive advantage."

11.4.2 Examples of Digitally Impacted Competencies

11.4.2.1 Competencies in Detached Global Telework

Individual service providers using global digital online platforms can be characterized as global teleworkers. It can be argued that remote workers such as freelancers doing their job from home or café may need less socializing initially than would a typical worker to be functional off-site. There is also some evidence suggesting that teleworkers value autonomy more than other workers do, and that they resist socializing except when it coincides with personal values related to work achievement [56]. Lamond [57] speculates that in teleworking, the importance of personality might be intensified because feedback and correction processes are not present in the same way as for office workers. Thus, they consider conscientiousness and personal competencies as characteristics, and interpersonal and technical competences as operative actions helpful for teleworkers. However, they note that there has been very little systematic research examining the effects of different teleworking practices and personality dimensions or competencies on the performance of teleworkers, or other indicators of successful adaptation. It seems that the key characteristics of teleworkers revolve around autonomy and self-leadership, i.e., the ability to work independently (in isolation), motivate oneself, self-discipline, and so on. Therefore, individual mental space seems to be a resource for working. The skillful operative actions that seem essential are communication skills, managing one's time, and being proficient with information and communication technologies, i.e., the virtual and social spaces that are used. It might be that these types of competencies are also needed in freelance work in digital global environments.

11.4.2.2 Competencies in 3D Printing

Replacing the traditional spare parts production with 3D technology is a massive change for organizations and their personnel. The whole process of spare part production from customer order to delivery needs to be reconsidered (Fig. 11.3). This means eliminating some functions (e.g., storing), creating new tasks (e.g., 3D modeling), and collaboration networks (e.g., new subcontractors) as well as setting up new management systems and evaluation criteria. As far as we know, the competencies needed in 3D printing have not been systematically analyzed before. Studying them is, however, crucially important for both theoretical and practical reasons. Theoretically, it is important to scrutinize and forecast the talent-related changes in the working life. Practically, this kind of analysis is a precondition for individual organizations in acquiring the skills needed for future success.

The data for the research was gathered by interviewing 11 professionals who were either already utilizing 3D printing or considered utilizing it in spare part production. The semistructured interviews lasted about 1 hour. The interviews were transcribed and analyzed using ATLAS.ti according to grounded theory and open coding strategies. In interviews, interviewees were asked about competencies "needed" when using 3D printing for spare part production.

The competencies needed for 3D printing of spare parts can be broadly categorized into three categories:

- *3D printing as a process* is a large category and encompasses skills and behaviors tightly related to 3D printing as a process. It consists of eight subcategories: (1) *general know-how of the 3D printing as a technology* (e.g., understanding what can and cannot be 3D printed, knowledge of different 3D printing techniques, and the added value of 3D printing), (2) *understanding the features of a spare part to be 3D printed* (use, special requirements), (3) *designing skills* (expertise in designing and scanning 3D models, software skills, and capability to interpret 3D models), (4) *skills to manage data and information security* (sending and handling 3D model files securely), (5) *knowledge of materials and material optimizing* (recognizing material characteristics, knowing requirements for materials, statics, metallurgy), (6) *3D printing skills* (mastering different printing methods, printing machines, ability to maintain the machines), (7) *processing and machining skills* (polishing, painting, etc.), and (8) *quality checking skills* (testing the printed products and verifying the required features).

- *Commercial and business skills.* It includes p*roject management skills, logistics skills, managing delivery times, general know-how on the law (particularly NDAs), and marketing as well as pricing skills.*
- *People skills*, or (1) *collaboration and communication skills* (both intra- and interorganizational), (2) *self-leadership skills* (ability to manage work and to develop oneself, positive attitude toward 3D), and (3) *ability to think creatively and with logical reasoning* (e.g., turning some weaknesses in the process into strengths).

The results imply that the computerization and digitalization of spare parts production radically change the competence needs. The importance of routine tasks seems to decrease (e.g., storing), while more demanding knowledge-based tasks (e.g., understanding the 3D printing technique) become more important. In addition, employees are strongly empowered and expected to independently self-manage their tasks. This empowerment is very descriptive of the postbureaucratic model of work—the model that seems to also get support from the present research.

11.5 Conclusions: Competencies for the Future

The key issue for the future is the resilience competency. Giustiniano et al. [11, 3] define resilience as "connoting capacities to absorb external shocks and to learn from them, while simultaneously preparing for and responding to external jolts, whether as organizations, teams, or individuals." Resilience is claimed to be necessary to protect actors and agencies from shocks, crises, scandals, and business fiascos that generate fear and create dissonance. Resilient people and organizations get knocked down and get up again, ready to learn from events and to be ready for future challenges. Individual, team, and organizational resilience are interdependent of each other. Building resilience on the individual level can spread within organizational settings and beyond, and collective cultural resilience can make individuals more resilient.

There have been several institutionalized efforts to identify the generic societal competencies needed now and for the future (e.g., [54, 58]). The European Commission prepared a European reference framework "*Key competence for lifelong learning*" considering competencies a combination of knowledge, skills, and attitudes. Key competencies are needed for personal development, employability, social inclusion, and active citizenship. Altogether eight generic key competencies were listed: literacy; languages; science, technological, engineering, and mathematical; digital; personal, social, and learning; civic; entrepreneurship; and cultural awareness and expression competencies.

In particular, the content and models for competencies of the digital workforce [59, 29] have been proposed as generic future competencies. For example, Colbert et al. [29] suggest considering how the competencies developed by digital natives and digital immigrants could benefit the organizations in which they work. They list some beneficial competencies gained from analyzing the present digital workforce:

1. *Digital fluency*. Proficiency and comfort in achieving desired outcomes using technology. Those who are digitally fluent have achieved a level of proficiency that allows them to manipulate information, construct ideas, and use technology to achieve strategic goals.
2. *Leadership skills*. Playing online games and interacting in virtual worlds can develop leadership skills that could transfer to work.
3. *Risk-taking*. Online games may also train people to task risks and learn from their mistakes, which is valuable in many of today's workplaces.

Carretero et al. [59] describe a digital competence framework developed as well by the European Commission. They focused on digital competency as the most critical among the eight key competencies for lifelong learning. Generally, Ferrari [60, 61] defines digital competency as "the set of knowledge, skills, attitudes, abilities, strategies, and awareness that are required when using ICT (information and communication technologies) and digital media to perform tasks; solve problems; communicate; manage information; collaborate; create and share content; and build knowledge effectively, efficiently, appropriately, critically, creatively, autonomously, flexibly, ethically, reflectively for work, leisure, participation, learning and socializing." The proposed digital competence framework consists of 21 competences divided into five areas [60, 5–7]:

1. *Information*: to identify, locate, retrieve, store, organize, and analyze digital information, judging its relevance and purpose.
2. *Communication*: to communicate in digital environments, share resources through online tools, link with others and collaborate through digital tools, interact with and participate in communities and networks, and cross-cultural awareness.
3. *Content creation*: to create and edit new content (from word processing to images and video); integrate and reelaborate previous knowledge and content; produce creative expressions, media outputs, and programming to deal with and apply intellectual property rights and licenses.
4. *Safety*: personal protection, data protection, digital identity protection, security measures, safe and sustainable use.
5. *Problem-solving*: to identify digital needs and resources, make informed decisions on the most appropriate digital tools according to the purpose or need, solve conceptual problems through digital means, creatively use technologies, solve technical problems, and update one's own and others' competence.

The reflection of the fifth surge of the "age of information and communications" on the organizational level has been a rather slow transformation. First, it resulted in virtual organizations, projects, and teams in the 1990s [62]. Today, organizations use virtual teams and projects as their common practice. In the 2000s, smart mobile technologies and wireless connections enabled work to be independent of time and place [1], even more so on the individual level. Locally, the number of mobile and multilocal digitally supported employees using several places to work and collaborate outside their main office started to increase, and employees became more detached from their mother organizations than before. This has indicated the need for more autonomous and self-regulating employees than before.

The 2010s brought the development to the global context; and doing delocalized microwork detached globally from local institutions started to grow. These workers can be characterized as a special type of remote worker—"detached temporary global teleworkers." A whole working class of digital employees has emerged worldwide, working as service providers for international employers [38]. In digital work, social interaction fully occurs in a virtual space, making the global Internet a work platform in addition to a local working location. From the work organization viewpoint, a virtual space for work is a necessity for remote workers to access knowledge and their clients and to collaborate with colleagues.

Studies on both remote telework and dispersed virtual teams underline communication skills as the key element of a competency. A study [63] showed that communication competency differs significantly in face-to-face and computer-mediated text-based communication, showing clear differences in several components: motivation, knowledge, attentiveness, expressiveness, composure, and coordination. Hertel et al. [48], in their study on virtual teamwork, state that in telecooperation the following are relevant: self-management skills, interpersonal trust, and intercultural skills. Studies on competencies needed in global virtual teams may help to ascertain what is needed by global digital labor and detached global teleworkers. They are often doing microtasks on a freelance basis, although more complicated tasks such as programming and design are increasingly used. Microwork on platforms is an extreme example of standardized and delocalized knowledge work. It is also a particularly good example of placeless work; it can, and indeed must, be performed without access to a physical worksite, and in principle could be performed from anywhere in the world with an Internet connection. In many cases, microwork's placelessness is produced intentionally, with a view to exploiting geographic differences in skills and labor costs, or the power that the ability to selectively compress time and space brings to its wielder.

The COVID-19 epidemic created a natural experiment that highlighted the importance of competencies to adapt and overcome the abrupt changes in work and its contexts. A survey by EuroFound[4] in April 2020 shows that over a third (37%) of those working in the EU began to telework as a result of the pandemic—over 30% in most member states. The largest proportions of workers who switched to working from home were

---

[4]https://www.eurofound.europa.eu/publications/report/2020/living-working-and-covid-19-first-findings-april-2020

found in the Nordic and Benelux countries (close to 60% in Finland and above 50% in Luxembourg, the Netherlands, Belgium, and Denmark, and 40% or more in Ireland, Sweden, Austria, and Italy). For comparison, a present US survey study[5] collecting a total of 25,000 responses from April 1 until April 5, 2020, showed that over one-third of workers responded to the pandemic by shifting to remote work, while another 11% were laid off or furloughed. Younger people were more likely than older people to switch from commuting to remote work. A representative survey among the German working population[6] tells that every second one was working fully or partly from home in mid-March 2020. Meanwhile, 41% responded that their job is not suitable for working from home. The studies on telework during the pandemic show that most teleworkers collaborated virtually with their colleagues, managers, and customers during their working days. Therefore, digital competences are especially needed in remote virtual work. These and similar studies around the globe tell that this "natural experiment" has until now brought forth unanswered questions on how to anticipate these kinds of partly unexpected situations, organize remote work and working conditions, and needed social and virtual support.

Finally, what is digital work and its need for competencies? It is evident that digitalization changes in many ways our working environment, work processes, task and job content, structures and organizations, and products and services—resulting in the need for partly and completely new competencies. On the societal level, the transformation of the competency need does not happen abruptly and accidentally. *Many of the changes in work processes are evolutionary rather than revolutionary* when digital solutions and innovations are crawling rather slowly into different components of the work systems and their environments. This development results in various types of present and future jobs—some are hybrid and some others are completely new. Their common feature is the multipurpose use of digital technologies, especially those technologies used for communication and collaboration and the search for new knowledge.

---

[5]Erik Brynjolfsson, John Horton, Adam Ozimek, Daniel Rock, Garima Sharma, Hong Yi Tu, April 8, 2020, COVID-19 and Remote Work: An Early Look at US Data. https://john-joseph-horton.com/papers/remote_work.pdf

[6]https://www.bitkom.org/Presse/Presseinformation/Corona-Pandemie-Arbeit-im-Homeoffice-nimmt-deutlich-zu

Acknowledgments

This chapter has been produced with the support of the Finnish Work Environment Fund for the project 'Limits and opportunities of autonomy in micro- and small- and medium-sized companies (AURA)'.

References

[1] Andriessen, J. H. E and Vartiainen, M. (eds.) (2006). "Mobile Virtual Work: A New Paradigm?" Heidelberg: Springer.

[2] Eurofound (2020a). "Telework and ICT-based mobile work: Flexible working in the digital age". Luxembourg: New forms of employment series, Publications Office of the European Union.

[3] Koroma, J., Hyrkkänen, U. and Vartiainen, M. (2014). "Looking for people, places and connections: hindrances when working in multiple locations - A review" New Technology, Work and Employment, 29(2), 139-159.

[4] Frey, C. B. and Osborne, M. A. (2013). "The future of employment: How susceptible are jobs to computerisation?" OMS Working Papers, September 18. http://www.futuretech.ox.ac.uk/sites/futuretech.ox.ac.uk/files/The_Future_of_Employment_OMS_Working_Paper_0.pdf

[5] Howcroft, D. and Bergvall-Kåreborn, B. (2019). "A typology of crowdwork platforms". Work, Employment and Society, 33(1), 21-38.

[6] Meil, P. and Kirov, V. (eds.) (2017). "Policy implications of virtual work". Switzerland: Palgrave Macmillan.

[7] Lehdonvirta, V. (2018). "Flexibility in the gig economy: managing time on three online piecework platforms". New Technology, Work and Employment, 33(1), 13-29.

[8] Lehdonvirta, V. (2016). "Algorithms that divide and unite: Delocalisation, identity and collective action in 'microwork'". In Flecker, J. (ed.) Space, Place and Global Digital Work, pp. 53-80. UK: Palgrave Macmillan.

[9] Wiesenfeld, B. M., Raghurum, S. and Garud, R. (1999). "Managers in a virtual context: The experience of self-threat and its effects on virtual work organizations". In C. L. Cooper and D. M. Rousseau (eds.), The Virtual Organization. Trends in Organizational Behavior, Vol. 6, pp. 31-44. Chichester: John Wiley & Sons.

[10] Lahti, E. (2019). "Embodied fortitude: an introduction to the Finnish construct of sisu". International Journal of Wellbeing, 9(1), 61-82.

[11] Giustiniano, L., Clegg, S. R., Cunha, M. P. and Rego, A. (2018). "Theories of organizational resilience". Cheltenham, UK: Edward Elgar Publishing.

[12] Méda, D. (2019). "Three scenarios for the future of work". International Labour Review, 158(4), 627-652.

[13] Kondratieff, N. D. (1979). "The long waves in economic life". Review (Fernand Braudel Center), 2(4), 519-562. Retrieved from http://www.jstor.org/stable/40240816

[14] Perez, C. (2014). "Financial bubbles, crises and the role of government in unleashing golden ages". In A. Pyka and H-P. Burghof (eds.), Innovation and Finance (pp. 11-25). London: Routledge.

[15] Wilenius, M. and Kurki, S. (2012). "Surfing the sixth wave. Exploring the next 40 years of global change". Finland Futures Research Centre, FFRC eBook 10. Turku: University of Turku.

[16] Dahlin, K. B. and Behrens, D. M. (2005). "When is an invention really radical? Defining and measuring technological radicalness". Research Policy, 34(5), 717-737.

[17] Linturi, R. and Kuusi, O. (2018). "Societal transformation 2018–2037: 100 anticipated radical technologies, 20 regimes, case Finland". Helsinki, Parliament of Finland, Committee for the Future, 2019. 485 s. Publication of the Committee for the Future 10/2018.

[18] Eurofound (2018). "Automation, digitalisation and platforms: Implications for work and employment". Luxembourg: Publications Office of the European Union.

[19] Moore, G. E. (1965). "Cramming more components onto integrated circuits". Electronics, 38(8), 114-117.

[20] Moore, G. E. (1975). "Progress in digital integrated electronics". Technical Digest 1975. International Electron Devices Meeting, IEEE, 1975, pp. 11-13.

[21] Eurofound (2020b). "Game-changing technologies: Transforming production and employment in Europe". Luxembourg: Publications Office of the European Union.

[22] Pajarinen, M., Rouvinen, P. and Ekeland, A. (2015). "Computerization Threatens One-Third of Finnish and Norwegian Employment". ETLA Brief No 34. http://pub.etla.fi/ETLA-Muistio-Brief-34.pdf

[23] Frei, F., Hugentobler, M., Schurman, D., Duell, W. and Alioth, A. (1993). "Work design for the competent organization". Westport, CT: Quorum Books.

[24] Riso, S. (2019). "Mapping the contours of the platform economy". Working Paper. European Foundation for the Improvement of Living and Working Conditions (Eurofound). Eurofound reference number: WPEF19060.

[25] Mosco, V. (2014). "To the cloud. Big data in a turbulent world". Boulder, CO: Paradigm Publishers.

[26] The IBM Institute for Business Value (2014). "The New Software Defined Supply Chain: Preparing for the disruptive transformation of Electronics design and manufacturing".

[27] Brynjolfsson, E. and McAfee, A. (2011). "Race against the machine: How the digital revolution is accelerating innovation, driving productivity, and irreversibly transforming employment and the economy". Lexington, MA: Digital Frontier Press.

[28] Rani, U. and Grimshaw, D. (2019). "Introduction: What does the future promise for work, employment and society?" International Labour Review, 158(4), 577-592.

[29] Colbert, A., Yee, N. and George, G. (2016). "The digital workforce and the workplace of future". Academy of Management Journal, 59(3), 731-739.

[30] Spencer, B. (2013). "Mobile users can't leave their phone alone for six minutes and check it up to 150 times a day". Daily Mail, 10 February. Available at http://www. dailymail.co.uk/news/article-2276752/Mobile-users-leave-phone-minutes-check-150-times-day.html.   [accessed October 12, 2017].

[31] Kelliher, C. and Richardson, J. (eds.) (2011). "New Ways of Organizing Work. Developments, Perspectives, and Experiences". New York: Routledge.

[32] Baffours, G. G. and Betsey, C. L. (2000). "Human resources management and development in the telework environment". Paper prepared for U.S. Department of Labor Symposium, "Telework and the New Workplace of the 21st Century," New Orleans, LA, October 16, 2000.

[33] Hislop, D. and Axtell, C. (2009). "To infinity and beyond?: workspace and the multi-location worker". New Technology, Work and Employment, 24(1), 60-75.

[34] Lehdonvirta, V., Kässi, O., Hjorth, I., Barnard, H. and Graham, M. (2019). "The global platform economy: a new offshoring institution enabling emerging-economy microproviders". Journal of Management, 45(2), 567-599.

[35] O'Leary, M., Orlikowski, W. and Yates, J. (2002). "Distributed work over the centuries: Trust and control in the Hudson's Bay Company", 1670-1826. In P. J. Hinds and S. Kiesler (eds.), Distributed Work (pp. 27-54). Cambridge, MA: The MIT Press.

[36] Hinds, P. J. and Kiesler, S. (eds.) (2002). "Distributed Work". Cambridge, MA: The MIT Press.

[37] Cohen, S. G. and Gibson, C. B. (eds.) (2003). "Virtual Teams that Work: Creating Conditions for Virtual Team Effectiveness". San Francisco, CA: Jossey-Bass.

[38] Scholz, T. (ed.) (2013). "Digital Labor. The Internet as Playground and Factory". New York: Routledge.

[39] Kellogg, K. C., Valentine, M. A. and Christin, A. (2020). "Algorithms at work: the new contested terrain of control". Academy of Management Annals, 14(1), 366-410.

[40] Rosenblat, A. and Stark, L. (2016). "Algorithmic labor and information asymmetries: a case study of Uber's drivers". International Journal of Communication, 10, 3758-3784.

[41] Lehdonvirta, V. (2017). "Where are online workers located? The international division of digital gig work". Blog accessed 8th October 2017 https://www.oii.ox.ac.uk/blog/where-are-online-workers-located-the-international-division-of-digital-gig-work/

[42] Stevens, G. W. (2013). "A critical review of the science and practice of competency modelling". Human Resource Development Review, 12(1), 86-107.

[43] Teodorescu, T. (2006). "Competence versus competency. What is the difference?" Performance Improvement, 45(10), 27-30.

[44] Spencer, L. M. and Spencer, S. M. (1993). "Competence at Work. Models for Superior Performance". New York: John Wiley & Sons.

[45] Boyatzis, R. E. (1982). "The Competent Manager. A Model for Effective Performance". New York: John Wiley & Sons.

[46] Bartram, D. and Roe, R. A. (2005). "Definition and assessment of competences in the context of the European diploma in psychology". European Psychologist, 10(2), 93-102.

[47] Roe, R. A. (2002). "What makes a competent psychologist?" European Psychologist, 7(3), 192-202.

[48] Hertel, G., Konradt, U. and Voss, K. (2006). "Competencies for virtual teamwork: development and validation of a web-based selection tool for members of distributed teams". European Journal of Work and Organizational Psychology, 15(4), 477-504.

[49] Hansson, H. (1998). "Kollektiv kompetens". Göteborg: BAS (in Swedish).

[50] Sandberg, J. and Targama, A. (1998). "Ledning och förståelse". Lund: Studentlitteratur.

[51] Hamel, G. (1994). "The concept of core competence". In Hamel, G. and Heene, A. (eds.), Competence-based Competition, pp. 11-33, Chichester: John Wiley & Sons.

[52] Javidan, M. (1998). "Core competence: what does it mean in practice?" Long Range Planning, 31(1), 60-71.

[53] Prahalad, C. K. and Hamel, G. (1990). "The core competence of the corporation". Harvard Business Review, May–June, 79-91.

[54] European Commission (2018). "Proposal for a COUNCIL RECOMMENDATION on Key Competences for LifeLong Learning". {COM(2018) 24 final}. Commission staff working document, Brussels, 17.1.2018. Retrieved 11.3.2019 from https://ec.europa.eu/transparency/regdoc/rep/1/2018/EN/COM-2018-24-F1-EN-MAIN-PART-1.PDF

[55] Athey, T. R. and Orth, M. S. (2009). "Emerging competency methods for the future". Human Resource Management, 38(3), 215-226.

[56] Omari, M. and Standen, P. (1996). "The impact of home-based work on organisational outcomes". Paper presented at the Australian and New Zealand Academy of Management Conference, Wollongong, NSW, 4–7 December.

[57] Lamond, D. (2003). "Teleworking and virtual organisations: The human impact". In D. Holman, D., Wall, T. D., Clegg, C. W., Sparrow, P. R. and Howard, A. (eds.), The New Workplace: A Guide to the Human Impact of Modern Working Practices, pp. 197-218. Chichester, UK: Wiley.

[58] UK Commission for Employment and Skills (2014). "The Future of Work: Jobs and skills in 2030". Evidence Report 84, February 2014.

[59] Carretero, S., Vuorikari, R. and Punie, Y. (2017). "DigComp 2.1: The digital competence framework for citizens with eight proficiency levels and examples of use". EUR 28558 EN.

[60] Ferrari, A., Brečko, B. N. and Yves Punie, Y. (2014). "DIGCOMP: a framework for developing and understanding digital competence in Europe". eLearning Papers, www.openeducationeuropa.eu/en/elearning_papers, number 38, 1-14.

[61] Ferrari, A. (2012). "Digital Competence in Practice: An Analysis of Frameworks". Sevilla: Joint Research Centre (JRC), European Commission.

[62] Davidow, W. and Malone, M. (1992). "The Virtual Corporation". New York: Harper Business.

[63] Schulze, J., Schultze, M., West, S. G. and Stefan Krumm, S. (2017) "The knowledge, skills, abilities, and other characteristics required for face-to-face versus computer-mediated communication: similar or distinct constructs?" Business Psychology, 32, 283-300.

12

Dominant Technology and Organization: Impact of Digital Technology on Skills

Steven Dhondt[1], **Frans van der Zee**[2], **Paul Preenen**[2], **Karolus Kraan**[2], **and Peter R.A. Oeij**[2]

[1]KU Leuven, Leuven, Belgium; TNO, Netherlands Organisation for Applied Scientific Research, Leiden, The Netherlands
[2]TNO, Netherlands Organisation for Applied Scientific Research, Leiden, The Netherlands

Abstract

This chapter describes a new approach to investigate, unravel, and explain the implications of digital technologies for skills. To do so, the chapter develops an approach to assess technology in companies in a more precise way, building on three main arguments. Firstly, current approaches to the subject treat all (new and emerging) technologies as equal. A more specific approach to technology is needed. Secondly, instead of starting from the potential of digital technologies, the focus should be on how technology investment decisions of companies are actually taken. Companies do not automatically reason from the available technology potential, but rather build on their current technology and capital stock and competitive position (the potential of technology). Thirdly, the organizational context should be considered. The actual use of skills in companies is strongly related to the organizational context. This is identified as the dominant organizational context. Based on these three main arguments, a new framework for work and skills technology impact research is suggested. Subsequently, the framework is applied to two professions in Dutch industry.

Keywords: Skill development, future of manufacturing, Industry 4.0, work design, digital skills

12.1 Introduction

The increasing use of advanced digital technologies is transforming innovation and production activities [1]. This also changes the requirements for skills within and between organizations, sectors, and countries, and may even render existing skills redundant or outdated [2, 3, 4, 5]. The new digital paradigms include a wide range of enabling technologies, such as the Internet of Things, additive production, big data, artificial intelligence, cloud computing, and augmented and virtual reality [6]. However, although substantial research is being conducted into the relationship between digital innovation and skills, especially also in relation to ICT-acceptance (e.g., [7]), the impact of new digital paradigms for innovation, production, and skills needs to be further explored [8].

The existing research on the impact of new (digital) technologies on labor and skills is mainly focused on the impact of (digital) technology on the number and type of jobs [9, 10, 11]. Yet, the predictions in these studies are rather mixed. Some studies predict that between 40 and 90% of jobs maybe lost [12], while in other studies, these percentages are around 10% [13]. This often gives too static an interpretation of what professions are. Little or no account is taken to mitigating effects such as the job content changes because of technology. New technology creates new types of jobs and new technology can lead to more work if the demand for products increases [13]. In fact, the introduction of these types of technologies is also much slower than expected [14]. In practice, the range of tasks in professions appears to be much broader and more adaptable than expected—which means that professions continue to exist, even if many tasks are computerized [9]. In general, the research has also mainly been approached from a macro and policy perspective (e.g., [15]) and based on existing but limitative datasets [10, 12]. To the best of our knowledge, no new data or monitors have been developed to assess the impact of new technology.

Another important criticism of existing research is that technology has only been operationalized in a limited fashion (e.g., [13, 10, 12, 16]). Frey and Osborne [12], e.g., only look at the possibilities of artificial intelligence (AI) and robots to overcome bottlenecks in professions. The most contentious point in their approach, however, is that computerization risk is measured by looking at the degree jobs contain certain skills: social, communicative, mathematical, problem solving, and ICT skills. The degree current jobs contain these skills predicts the degree that they will be computerized. It should be clear that predicting future skill needs with such an approach will lead to tautological reasoning. Graetz and Michaels [17] and Acemoglu and Restrepo [18, 19, 20] look at robot technology present in a region. Bessen et al. [21] only focus on investments in technology. Brynjolfsson et al. [22] also focus on a limited number of digital technologies, in particular AI, but do not empirically investigate this.

And finally, an important point of criticism is that the organizational context must be considered but is generally ignored in the analysis [23]. There is a lack of specific insights into how specific digital technologies are implemented and developed in organizations. In most research, this organizational dimension is a black box, as it is completely absent [including 12, 16, 13]. However, these insights into the relationship between technology and organization are crucial to understand and explain the real impact of digital technologies on labor, organizations, tasks, and skills.

Therefore, the understanding of the impact of digital technology needs further development. Technology must better be operationalized, and the organizational context and aspects of work, as well as technology investment factors should be included in future research as well. This chapter introduces two core concepts for research into the impact of digital technologies on work in companies and sectors that address the above issues and can bring the discussion further: the dominant technological context and the dominant organizational context. These concepts will be operationalized and the benefits of using these concepts in research will be shown. An example of changes in skill demand within and between jobs in Dutch manufacturing industry shows the strength of this approach.

12.2 Dominant Technology

To examine the impact of new digital technologies on skills and organizations, it is important to determine what the dominant technologies will be for organizations. We believe that three elements are important for this: first, specify the focus technologies; second, understand the heterogeneity of technology in organizations, i.e., vintage and investments in technology; and, third, the measurement of technology. These elements are explained hereunder. The ideas are applied to the example of the Dutch industry.

12.2.1 Five Technology Types

In the current research into digital technology, the implicit assumption is that it does not matter which manifestations technology takes. In the recent

World Economic Forum report 2018 [24], technology was conceived as the degree of adoption of 19 specific technologies (from big data analytics to drones). The technology adoption rate is the percentage of companies that have implemented the technology. Bessen et al. [21] (2019) see automation as the cost of automation. They make the distinction between automation costs and computer investments. However, they mainly gain insight into technology investments rather than the existing technological situation in a company. Also, the Internet of Things, additive manufacturing, big data, AI, cloud computing, augmented and virtual reality, cobots, etc. are all assumed to have the same impact on the scope and quality of the work. In other examples of this approach, digital technology is sometimes measured using proxy variables, such as the percentage of robots in a country [17] or calculated at the level of a region [18, 19, 20, 25]. Some authors make more distinctions between types of technology. Acemoglu and Restropu [18, 19, 20] distinguish between two types of capital: low-skilled automation and highly-skilled automation. They link these differences in automation to examples such as "industrial robots" and "AI." The Acemoglu and Restropu model does not consider other types of technology.

From an organizational point of view, not all technologies are the same and they not only influence the complexity of tasks, but also the way in which organizations are managed. New technology thus influences various aspects of organizing that determine productivity. Bloom et al. [26] have proposed another classification. They assert that ICT should not be viewed as a whole, but that information technology and communication technology, in particular, have different organizational consequences. Information technology (e.g., ERP, CAD/CAM) helps to strengthen the ability of middle managers to search for solutions, so that they can be expected to broaden their jobs and grow in autonomy and decentralization ("data access": [27]). Communication technology, on the other hand (e.g., e-mail and communication networks), ensures that decision-making and coordination can take place more quickly. This allows middle managers to specialize more in what they are strong at and allows central managers to more quickly ensure alignment between middle managers. Communication technology therefore leads to task specialization and centralization of decisions. In fact, their distinction can also be used for the relationship between (all) management levels and the first-line workers. Communication technology affects the relationships from top to bottom in an organization; information technology allows for task enlargement at all levels. Ter Weel [28] adds to this distinction that technologies can be aimed at automating tasks, or at increasing the capacities of employees (in line with information technology). We would like to add that innovations are also possible in management systems or organizational measures [29, 30, 31]. These different technologies can be more or less aligned in their implementation in organizational settings. Innovations in other dimensions of technology can be in line with impacts of information and communication technology.

With these distinctions, the complex technological developments can be reduced to the five categories: information technology, communication technology, management systems, "hard" automation, and human enhancement technology. These technologies have distinct predicted impacts on employment dimensions. Table 12.1 provides an overview of these five types of technology, with their process impacts and expected labor impacts.

12.2.2 The Process of Technology Implementation: Vintage and Investments

In addition to the distinction between types of technology, the process aspect in technology needs to be considered. As was indicated, the potential of technology, the investment strategy, and the available technology differ by nature. The process perspective helps to make understand the heterogeneity of the technology situation in organizations.

A starting point to understand the technological situation in organizations is to find out what is the current technology base of a company. Porter [32] saw technology as part of the core competence(s) of a company. The success of the application of new digital technologies depends on the current technological knowledge in the company. In the 1980s, there was a great deal of attention for the composition of the capital stock of companies. In calculations of total factor productivity, an estimate of the capital stock is needed. In this context, much has been done about the calculation and composition of (company) capital. The reasoning is that the strength of a company depends on the extent to which the capital consists of new "vintages" of technology. Old vintages of capital must be replaced by new ones to cope with rising wages [33]. Dutch economists have long time discussed the importance of vintage effects of capital during the 1980s [34]. The opinion was that the low wage policy followed by the Dutch governments removed an incentive to invest in new technology, with the result that productivity fell due to outdated technology ("old vintages"). A criticism of this view is that vintage only "looks back," i.e., determines what

Table 12.1   Five technology types and their impact.

Type of technology	Process impact	Employment impact
Hard automation	Technology can automate human labor. This usually involves "hard automation" in which technology takes over the actions and tasks of employees completely. Examples are robots or automatic welding machines. Other examples in logistics are self-driving cars and trains, autonomous ships. It is important in hard automation that it does not always involve equipment. Sometimes it is possible to have customers do more work so that work can disappear. Think of supermarkets where cashiers become redundant when customers do self-scanning.	Disappearing tasks and occupations
Human enhancement (supporting) technology	Technology can support employees in the execution of their tasks. This usually involves mechanical tools, but it can also involve automated tools such as exoskeletons (in construction and assembly) or digital tools such as vision picking (in logistics).	Enlargement of operator capabilities. Increased productivity.
Communication technology	This technology focuses on the communication processes between employees or between employees and managers. Communication with the outside world (e.g., customers, suppliers) can also be structured differently using this technology. Communication technology has contributed to the strong rise of global value chains. This technology fits in with the management processes in organizations. Communication technology does not always have to be mobile technology. A conveyor belt can also be seen as communication technology. Such a conveyor belt helps employees to specialize and centralizes decisions about the production process.	Strengthening of hierarchy, narrowing of tasks/specialization
Information technology	This is a separate technological change that fits in with the way in which employees gain access to information. Data access technology is mentioned in the literature: information technology helps to speed up access to information.	Decentralization, broadening of tasks
Management systems	With this technology, activities in organizations are to a large extent standardized and uniformed. Other technologies can play a role in such systems, but not necessarily. An example is the Lean Production system.	Quality improvement, productivity improvement, integration and specialization

existing technology is. However, recent investments in new technology can be an obstacle to investing in the most recent disruptive technologies. And new technologies may not always be the most efficient. "The results show that in some circumstances older vintages might appear on the efficiency frontier, unlike some newer vintages that are found to be inefficient, despite benefiting from the advancement of the technology" [35: 244].

There are few data sources available for the operationalization of "vintage." Calculating the impact of vintage is usually fairly complicated and the results only apply when all kinds of problematic assumptions are met. A recent German study, the IAB company panel survey [36], measures vintage in a qualitative way: companies indicate to what extent their technology is "out of date" or "state of the art." What is important about the vintage discussion is that new technology needs time to be implemented. It can take time before a new technology has a positive influence on production/service provision [37].

Secondly, it is also about the investments that companies make. What do companies want to change in their current capital? If the market offers a lot of "potential" of new technology, what exactly is the potential selected? If we take the results of Table 12.1 into account, can we then gauge what is being invested in? Traditionally, investments in "hard technology" are considered to show the capital investments of a country. For example, EU-KLEMS data[1] indicate how much was invested in tangible assets in the last year. Much is known about these investments at an aggregated level. Acemoglu and Restrepo [20] focus precisely on these investments. Two questions are important here. The first question is what exactly you want to know about this technology. The EU-KLEMS study is mainly concerned with estimating the differences in the distance to the technological frontier. This frontier is a measure of technological advantage, and traditionally the country that generates the highest total factor productivity with a technology is seen as the frontier. The EU-KLEMS data provide information on the differences in technology. The changes in investments indicate how countries want to make up for their shortfall (or lead) [38, p. 106].

But it is not just about "hard technology." Various studies, sponsored by the OECD, have clarified that companies also invest money in less "hard" technologies to increase their productivity [39, 40]. This concerns "knowledge-based capital (KBC)" that consists of investments in ICT, R&D, management training, organizational capital, etc. [41, 42]. According to a

---

[1] www.euklems.net/

research by Corrado et al. [39], investments in KBC contribute about 7% to 8% to the annual growth in labor productivity, which is slightly more than the contribution of investments in "hard capital." However, the research on KBC is progressing slowly. The main reason is that only register data is considered, only what is available about investments in companies. In the KBC, investments in management quality and company-related R&D account for the largest share [43]. For management quality, several measures are used [44]. In Saia et al. [45], this quality is related to the PIAAC scores. Andrews and Westmore [44] look at the extent to which professionals are supported for management positions. Earlier KBC studies looked at the size of the management consultancy sector. The importance of KBC is that it is seen as an important factor in promoting the diffusion of new technology [43].

The research into "intangibles" is also relevant to this chapter because it provides us with insight into the scale of ICT investments. The main objective of the research by the EU-KLEMS network is to establish why all investments in technology and ICT do not yet lead to a productivity leap in economies [46]. Byrne and Corrado [46] identify several reasons why the expected productivity leap is not visible, such as the fact that ICT costs are falling very sharply. More work is needed here because it remains difficult to determine the impact of ICT investments on growth. Various elements of ICT appear to be difficult to grasp in growth statistics. For many aspects of the investments, it will actually be necessary to look at the company level to determine to what companies pay attention in technology and capital stock development [21].

12.2.3 The Potential of Technology

Most analyses of technology are about the potential of technology, what technology can do, and how that potential will have consequences for work. Most consultants focus on this potential [24, 15]. In addition to the fact that no coherence is seen, differentiation is made between all these technologies. There is also no weighing of which technology will have the most impact. The OECD [40] has proposed the "technology burst" imagery to determine differences in the potential of new technology. A "burst" exists when the number of patents in a technology field increases sharply at a given moment. In that case you can expect that new applications will flood the markets in the short term. Patents may not be the best indicator of the potential of technology. It is important to bear in mind that only half of the patents are

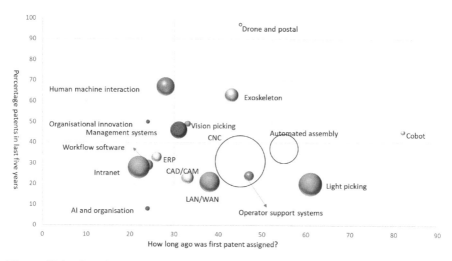

Figure 12.1 Overview of history of patent category, size of number of patents (size of bulb), and importance of number of patents in the past five years (Espacenet, https://nl.espacenet.com/ downloaded 10-8-2018). Legends: Black circle = technology that automates human labor ("hard technology");Blue = technology to support employees; Green = communication technology or technology that allows control from above; Orange/yellow = information technology that gives employees more access to information; and Red = management systems or organizational innovation.

used for a variety of reasons [38]. Despite this complication, a patent does give some indication of the development of the knowledge intensity of a sector and specialization of knowledge.

To illustrate this, we indicate what this "technology burst" reasoning yields if we use the distinction of five types of technology in the manufacturing industry. Using the Espacenet patent database, we made an inventory of patents related to the five technology types. Espacenet is a tool of the European Patent Office (EPO) and gives free access to more than 100 million patents from all over the world. For the purpose of our analysis, specific technology categories were investigated, the date on which we the first patent could be found, and the percentage of new patents for each category in the past five years. Figure 12.1 presents an overview of this analysis.

12.2.4 Measuring Dominant Technology

How can these ideas of type of technology, vintage, investments, and potential of technology be integrated to assess the technology situation in

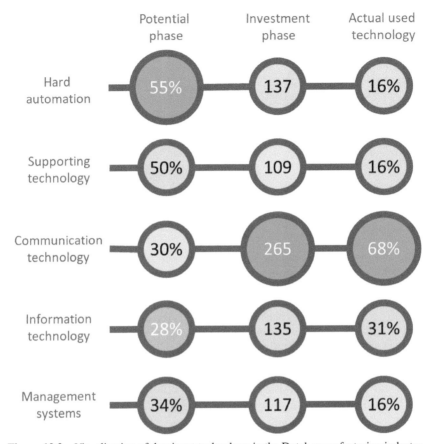

Figure 12.2   Visualization of dominant technology in the Dutch manufacturing industry.

an organization sector or country? This is where the concept of "dominant technology" becomes useful. It is appropriate to distinguish between the type of technology and the phase in the investment process, but how can one weigh the dominant technological position of a company? There are no approaches to this in the literature. We propose to develop a composite index to determine which technology is dominant. Figure 12.2 clarifies what we have in mind.

Figure 12.2 shows three phases in which a technology for a company, sector, or country is situated: the actual use, the investment phase, and the development or potential phase. The size of the bulb in the figure indicates how the technology in a phase relates to other technologies: the larger the bulb, the greater the importance of a technology in a phase. The colors of the

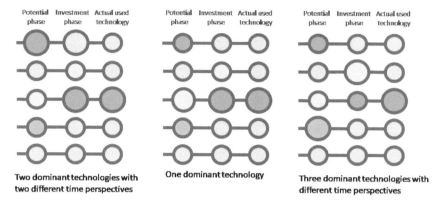

Figure 12.3 Examples of three alternative technology situations.

bulbs indicate the growth of the technology in a specific time context: dark green indicates strong growth in a technology, light green limited growth, and orange decline in growth. The figures in the bulbs mean the following:

- Actual technology use ("reality") is the percentage of companies that have implemented a specific technology.
- Investments ("strategy") are an index figure (with an index=100 for a comparable reference year) showing the growth compared to a reference year.
- Development ("potential") is the percentage of patents that have been approved in the past five years compared to the total number of patents in a technology.

Dominance is then determined as the technology that comes out strongest in the comparison between the three phases. The assumption is that the current reality must first be weighed, then the investments, and only then the potential phase. To determine, which technology will have the most influence on skills and work, we need to look not only at each phase, but also at all phases at the same time. This means that many combinations are possible, i.e., in the "vintage," e.g., communication technology can be dominant, and in the investment phase attention shifts to information technology. In the longer term, it maybe that the development in hard technology will weigh most heavily. Figure 12.3 shows three possible alternative technology situations.

In the left-hand side of Fig. 12.3, two technologies are dominant, but in different phases: hard automation is a future given, currently communication technology dominates the installed base of technology and the investments.

The centermost figure only shows one dominant technology: communication technology. The figure on the right shows a very dispersed picture: information technology promises a lot of changes in the future, supporting technology is most important in the investments, and communication technology dominates the installed base. The dominant technology situations may therefore sometimes be very clear, sometimes more heterogeneous.

This perspective means that the supply of technology itself is not enough to grasp what will happen in companies. First it is important to examine what technology is present in the companies and then in which investments are made. Only then is it useful to look at the supply of new technology.

12.2.5 Dominant Technology in the Dutch Manufacturing Industry

To illustrate the power of this approach, an example of the dominant technology in the Dutch manufacturing industry is developed. To determine the dominant technology, a combination of several information sources was used (see for full information [47]). For the existing technology, an analysis was made of the Dutch statistics data on ICT (*Statline data on ICT use by companies*) and the Netherlands Employers Surveys [48]. To calculate the investments in knowledge-based capital in the Dutch industry, some of the necessary data is available. The EU-KLEMS data indicate how much was invested in tangible and intangible assets in the past years. However, the intangible assets are not complete as described by Corrado et al. [39] and the OECD [40]. Costs incurred for HRM, e.g., organizational investments, are not included in the figures for EU-KLEMS. Corrado et al. limit themselves to R&D and software investments. In the Dutch manufacturing industry, the ratio between tangibles and intangibles has shifted in favor of intangibles in the short term. In recent years, companies have again started to invest more in "hard technology," but intangible investments appear to receive even more attention in the manufacturing industry in the past few years. If one looks at the industry (industry plus other sectors), the same increase can be seen in intangibles, thanks to the fact that hard investments still weigh more heavily.

Figure 12.2 shows the outcome of our analysis for the dominant technology in the Dutch manufacturing. The development in a technology is reflected in the series of three bubbles. The numbers in the bubbles represent the proportions between the bubbles at one particular moment:

- Potential: to assess the potential of technologies we use the percentage of patents in the last five years of total patents in a technology category in an industry (e.g., 55% = 55% of patents for hard automation have been issued in the last five years).
- Strategy: the investment index for 2011–2015 has been calculated, which reflects the growth in investment in a technology (2011=100) (e.g., investments in hard technology have increased by 37% during this period).
- Reality: the percentage of companies that have applied a technology from that category (e.g., 16% of companies have robots).

The size of a bubble indicates in which development perspective (potential, strategy, reality) a technology has the most weight. The largest bubble in a phase carries the greatest weight. The colors represent the situation in 2011–2017: dark green shows that there is strong growth in a technology; light green indicates that growth is limited; and orange indicates that there is decline. In the Dutch manufacturing industry, this means that communication technology can be expected to have the most important influence now and in the short term. This means that network technology, e-mail systems, and mobile technology will have the strongest impact on organizational processes and occupations. This technology strengthens the communication flows in the companies and, according to Bloom et al. [26], should result in a stronger centralization of decisions toward top management and a narrowing of occupations on the shop floor. In the longer term, "hard automation" in the Dutch industry seems to be bringing about major changes.

12.3 Dominant Organizational Context

12.3.1 Distinct Organizational Concepts

A second major determinant when looking at the impact of technologies on jobs and skills is the organizational context in companies and organizations. There is a general lack of insights into how specific digital technologies are implemented and developed in organizations. With management systems as 5th technology type, we already introduced at the dominant technology level an aspect of how work is organized. However, we need to clearly distinguish management systems from organizational concepts, even though in practice there maybe overlap between the two. Management systems are defined here as rules on how to work and the information systems that contain those rules. This means that these systems mainly include parts of the control system in an organization. Organizational concepts are broader and include the control, but also the views on how the work should be carried out, i.e., the division of labor. Organizational systems that are aimed at visualizing all aspects of the work are aimed at control. TQM, lean production, and other concepts try to operationalize quality and performance of production systems. Other concepts such as workplace innovation [31] or employee-driven innovation [49] are aimed at maximizing employee involvement to optimize innovation.

12.3.2 Measuring Organizational Concepts at Different Levels

The way in which technology is shaped within companies is also related to the organizational form. This presents a chicken-egg problem. According to economists [26, 50] organizational form is seen as an effect of the technology, whereas organizational sociologists indicate that the choice for the technology is an effect of targeted choices [51, 29, 30]. Changes in the organizational concept are regarded as an innovation instrument for companies. In the previous section, it has been indicated how companies have in recent years invested more in their organizational capital than in hard technology (see also [42]). The question is how the organizational concept itself can be operationalized. Different perspectives are discussed here. When looking at the organizational level, control and production processes should be considered. Control processes concern differences in the relationship between controlling, preparatory (planning), and supporting (maintenance) tasks. Decisions in these tasks can be centralized or concentrated. Centralization is about decisions in the execution of tasks in production or services, i.e., the distinction between managing and operational tasks. Concentration refers to bringing together preparatory and supporting tasks in production or service provision, i.e., creating complete or incomplete functions (job enrichment or the reverse) [52]. Companies can have all kinds of considerations to centralize or concentrate decisions. As the example of the Netherlands illustrates, in recent years, companies appear to have increasingly centralized their control systems, supported (or driven) by communication technology [53].

The introduction of this chapter indicated that insight can be gained into the organization of work at the individual level of a worker. The task analysis advocated in the British Skill Survey [54], the PIAAC [55], and recently in the Netherlands Skill Survey [56, 57] provide overviews of tasks related to preparation, support, and management within organizations. Occupations without such tasks seem to indicate a focus on operational tasks. Occupations that do contain such tasks operate in contexts in which employees themselves must control their environment and the content of the work. Lorenz and Valeyre [58] used the European Working Conditions Survey (Eurofound) to successfully distinguish, at the country level, the application of four organizational concepts: learning organizations, lean organizations, Taylorized organizations, and simple organizational forms. The "learning" model is characterized by a high degree of autonomy and task complexity, learning and problem solving, and a low degree of individual responsibility for quality management. The "lean" model is characterized by the presence of teamwork and job rotation, the variables for quality management, and the various factors that limit the pace of work. Job autonomy is relatively low and strict quantitative production standards are used to control employees' efforts. The "Taylorist" model shows minimal learning dynamics, low complexity, low autonomy, and an overrepresentation of the variables that measure the limitations of work pace. The "simple" model groups simple forms of work organization, where the methods are for the most part informal and uncodified. Advances in research into the impacts of technology will also have to take dominant organizational concepts into account in the research design.

12.3.3 Dominant Organizational Context in the Dutch Manufacturing Industry

What is then the current situation in the Dutch manufacturing industry? Dhondt et al. [59] investigated seven major occupational jobs using the Netherlands Skill Survey [60]. As explained earlier, the degree organizational tasks are included in a job, explains the degree of labor division in an occupation[2]. From the study, 33% of middle managers are operating in highly Taylorized work organizations. For packaging personnel, this percentage rises to 52%. The organizational concept used in a company is an important explanation of differences in occupational profiles. It is striking that in most occupational positions in the Dutch manufacturing, the degree of Taylorization plays an important role. The reported percentages have not changed in both groups in past years, but the expectation is that Taylorization in both occupations will increase. One explanation is that the dominant communication technology logic plays a role, because

---

[2]For more information on operationalisation, see Dhondt et al. [59]. A further elaboration of the results of the Netherlands Skill Survey is taken up in a follow-up article of this publication.

this technology ensures a stronger centralization of regulating tasks in organizations. Occupations will tend to specialize.

12.4 Predicting Impact of Dominant Technology and Organization on Skills

12.4.1 Future Impacts

The analyses in sections 2.4 and 3.3 indicate that communication technology and Taylorization will be the dominant technology and organization in the coming years. If this context remains dominant in the Dutch manufacturing, then further specialization at the occupational level and centralization of decision-making (with less autonomy for individuals) are the consequences. To illustrate this, results of the analysis of job profiles for middle managers and packaging jobs in the Dutch manufacturing are shown here. Both occupations are common in the manufacturing industry.

This chapter focuses on improving the discussion about the relationship between new technology and work. The core debate is about the relationship between technology and skills. Other dimensions (quality of work, wages, etc.) are equally important, but receive less attention in the debate. Acemoglu and Restrepo [20] and Bessen et al. [21], e.g., focus exclusively on skills development. Frey and Osborne [10] are interested in the size of employment in a job. Most of these studies assume a direct link between technology and labor aspects. Yet as indicated earlier, the organizational context must also be considered as a codeterminant of what kind of skills are needed. The skill debate revolves around two topics: the requested qualifications of individual workers and the issue of skills polarization at the workplace. The focus here is what impacts we can expect for the future of skills in the Dutch manufacturing industry, given the dominant technology and organization.

For estimating short-term impacts (three to five years), extrapolating trends still seems to be the most appropriate method. Other methods such as econometric estimations [12] or Markov chain analysis [16] are only fruitful for long-term forecasts. The extrapolation method involves assessing the development in a dimension of work and then estimating how the dominant technology and organization will influence the development in that dimension. Trend information in itself is insufficient. It is important to estimate, which factors will have more weight on the trend than others.

That is why it is important to consider "expert judgment" to weigh the development in the trend. If possible, the information should be enriched with calculated information (e.g., extrapolation of time series). The full prediction requires (a) the current trend development in an occupational dimension and (b) the trend extrapolation for that dimension to the short-term future. Other extrapolations are possible but make calculations unnecessarily complex. The advantage of the presented approach is that any deviation from the trend should be made explicit. We only speak of a trend if there is a statistically significant change (or continuity) in a dimension. This approach is illustrated hereunder.

12.4.2 Impact on Skills Within Jobs

Figures 12.4 and 12.5 show the differences in required qualifications in two occupations in the Dutch manufacturing industry—for middle managers and packaging jobs—considering different organizational contexts. Organizational context is operationalized at the individual level as described earlier.

Figures 12.4 and 12.5 show that both middle managers and packaging workers in a "full job function" are asked for significantly more social, communicative, STEM, and ICT competences than in a "Taylorized function." If we know that the future degree of Taylorization in the function will increase, then it is clear that in the long term, fewer social competencies

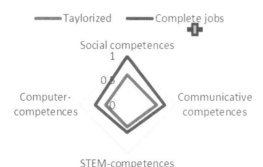

Figure 12.4 Middle managers: Overview of social, communication, STEM, and computer competencies present. Comparison between available competencies according to Tayloristic or complete working context. Differences have been tested for significance (+: p<0.05).

ISCO3 Packaging jobs (source: NSS2012-17)
(+: significant difference % present, p<0,05)

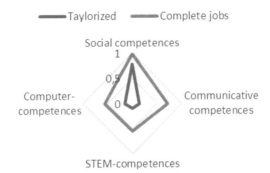

Figure 12.5   Packaging jobs: Overview of social, communication, STEM, and computer competences present. Comparison between available competencies according to Tayloristic or complete working context. Differences have been tested for significance (+: p<0.05).

will be required for middle managers and fewer requirements will be set for packaging workers at all skill demands. Tayloristic concepts always lead to less reliance on these skills than in organizational concepts with more complete occupations, regardless of the technology used. Probably the type of technology can help strengthen the choice for an organizational context and influence in that way certain impacts.

12.4.3 Impact on Skills Distribution Between Jobs

The second discussion is about polarization of skills at the workplace. Impacts of dominant technology and organization may not be visible within an occupation, but in the relationships between occupations. Therefore, at least a sector view of occupations is required. The changing relationships between jobs can be described as polarization. When looking at skills, the access to and the development of skills maybe very different between job categories. To illustrate the polarization effects, we look at the skill distance of the middle managers from the packers. To measure the distance, the level of higher skilled employees in a job is taken as an indicator: the percentage of bachelor or master's degree. In packaging jobs, this is 7%. In the period 2014–2017, however, this percentage has significantly risen. In the middle manager jobs, there are 2,4 times more highly educated persons.

The expectation is that the dominant technology and organization will not increase this skill polarization. The explanation is that the number of people with a bachelor or academic diploma among the middle managers has hit a limit: a higher percentage of high skilled managers seems improbable. If the degree of Taylorization increases, this will further reduce the skills required in both positions: Figures 12.4 and 12.5 show that more Taylorized positions make less use of the social, communicative, STEM, and computer skills. The polarization (i.e., the skill distance) may therefore decrease even more.

12.5 Conclusion and Discussion

This chapter proposes an alternative approach to analyzing the impact of digital technology on work in companies and sectors. In the literature, predictions about employment impacts are quite disperse. The main reason why predictions are still unreliable stems from the way in which technology, organization, and aspects of work are currently conceptualized. The example of the Dutch manufacturing industry shows that one can arrive at predictions that differ substantially from those by Frey and Osborne [10]. The example of the skill changes within and between jobs was used to illustrate the importance of a new approach to understanding technological and organizational change. The estimations for the Dutch manufacturing show that the skill requirements for jobs are very high in non-Taylorized environments, and quite low in Taylorized working environments. The expectation is that dominant technology, the stress on more communication technology, will lead to more Taylorized jobs, reducing the needs for the so-called 21st century skills. A second result is that there are polarized skill differences between job categories. These differences seem to become less in the future, because of the Taylorization. This is not itself a positive development for jobs, because the overall requirements are lower. Certain jobs will profit from more of these changes than the ones covered in this chapter.

New to this approach and monitoring system is the conceptualization of technology itself. In contrast to other approaches, a more "contained" approach to technology has been chosen in line with and following the process impact of technology: the impact is different when we talk about the potential of technology, the technology investment strategy, and the current technology stock. Five technologies have been mapped and described. The concept allows us to formulate theoretically substantiated conclusions about possible impact. Based on these conclusions, professions in sectors were examined. The model provides a broad impact framework, broader than currently used in the analysis of technological development. This chapter has limited itself to skill differences within jobs and between jobs. In Dhondt et al. [59], more work dimensions are used, also looking at the extent of employment in occupations, development in the quality of work, consequences for polarization in sectors, and perception of work dimensions. The model allows making predictions for the impact of technology on labor.

The task we have set ourselves is ambitious. More work is needed to elaborate and further operationalize the concepts developed in this chapter. Most work will be needed on the proposed three technology horizons. For the research strategy, combined data collection approach at the company and employee levels will be required. The bottom line is that the work on the future of technology and its impact is only starting. Better methods need to be developed. This chapter provides the first steps.

12.6 Acknowledgements

This research has been produced with the support of the Dutch Foundation Instituut GAK and the Horizon 2020 Beyond4.0 project (Grant Agreement: 822296; Deliverable 5.2). A previous version of this chapter was published in W. Bauer, Ganz, W. and Hamann, K. (eds.), International Perspectives and Research on the "Future of Work". International scientific symposium held in Stuttgart in July 2019. Stuttgart: Fraunhofer IAO, pp.186–206.

References

[1] Alcacer, J., Cantwell J. and Piscitello, L. (2016). "Internationalization in the information age: A new era for places, firms, and international business networks?" Journal of International Business Studies, 47(5), 499-512.

[2] Autor, D.H. (2015). "Why are there still so many jobs? The history and future of workplace automation". Journal of Economic Perspectives, 29(3), 3-30.

[3] Autor, D.H., Dorn, D. and Hanson, G.H. (2015). "Untangling trade and technology: Evidence from local labor markets". Economic Journal, 125(584), 621-646.

[4] Silva, H.C. and Lima, F. (2017). "Technology, employment and skills: A look into job duration". Research Policy, 46(8), 1519-1530.

[5] Zysman, J. and Kenney, M. (2018). "The next phase in the digital revolution: Abundant computing, platforms, growth, and employment". Communications of the Association of Computing Machinery, 61(2), 54-63.

[6] Rindfleisch, A., O'Hern, M. and Sachdev, V. (2017). "The digital revolution, 3D printing, and innovation as data". Journal of Product Innovation Management, 34(5), 681-690.

[7] Autor, D.H., Katz, L.F. and Krueger, A.B. (1998). "Computing inequality: Have computers changed the labor market?" Quarterly Journal of Economics, CXIII, 1169-1214.

[8] Consoli, D., Vona F. and Rentocchini F. (2016). "That was then, this is now: Skills and routinization in the 2000s". Industrial and Corporate Change, 25(5), 847-866.

[9] Atkinson, R.D. and Wu, J. (2017). "False Alarmism: Technological Disruption and the U.S. Labor Market, 1850-2015". Information Technology & Innovation Foundation ITIF. http://dx.doi.org/10.2139/ssrn.3066052.

[10] Frey, C.B. and Osborne, M.A. (2013). "The Future of Employment: How Susceptible are Jobs to Computerization?" Oxford: Oxford Martin Programme on Technology and Employment Oxford, University of Oxford.

[11] Van Roy, V., Vértesy, D. and Vivarelli, M. (2018). "Technology and employment: Mass unemployment or job creation? Empirical evidence from European patenting firms". Research Policy, 47(9), 1762-1776.

[12] Frey, C.B. and Osborne, M.A. (2017). "The future of employment: How susceptible are jobs to computerisation? Technological Forecasting and Social Change, 114, issue C, 254-280.

[13] Arntz, M., Gregory, T. and Zierahn, U. (2016). "The risk of automation for jobs in OECD countries: A comparative analysis". OECD Social, Employment and Migration Working Papers, No. 189, OECD Publishing, Paris.

[14] Van Helmond, C., Kok, R., Ligthart, P. and Vaessen, P. (2018). "Veel ruimte voor meer digitalisering Nederlandse maakbedrijven". ESB, 23(4), 2-3.

[15] McKinsey (2018). "Skill shift. Automation and the future of the workforce". Discussion Paper, McKinsey Global Institute, Brussels.

[16] Kim, Y.J., Kim, K. and Lee, S. (2017). "The rise of technological unemployment and its implications on the future macroeconomic landscape". Futures, 87, 1-9.

[17] Graetz, G. and Michaels, G. (2015). "Robots at work". CEP Discussion Paper No. 1335, Centre for Economic Performance, London School of Economics and Political Science, London.

[18] Acemoglu, D. and Restropo, P. (2017a). "Robots and jobs: Evidence from US labor markets". NBER Working Paper No. 23285, NBER, Cambridge.

[19] Acemoglu, D. and Restropo, P. (2017b). "The race between man and machine: Implications of technology for growth, factor shares and employment". NBER Working Paper No. 22252, NBER, Cambridge.

[20] Acemoglu, D. and Restropo, P. (2018). "Low skill and high skill automation". Journal of Human Capital, 12(2), 204-232.

[21] Bessen, J., Goos, M., Salomons, A. and Van den Berge, W. (2019). "Automatic reaction - What happens to workers at firms that automate?" Discussion Paper, CPB Netherlands Bureau for Economic Policy Analysis, Den Haag.

[22] Brynjolffson, E. and McAfee, A. (2014). "The Second Machine Age - Work, Progress, and Prosperity in a Time of Brilliant Technologies". New York: Norton.

[23] Agnew, A., Forrester, P., Hassard, J. and Procter, S. (1997). "Deskilling and reskilling within the labour process: The case of computer integrated manufacturing". International Journal of Production Economics, 52(3), 317-324.

[24] WEF (2018). "The future of jobs". Report 2018; Insight Report, World Economic Forum, Centre for the New Economy and Society, Cologny/Geneva.

[25] Dauth, W., Findeisen, S., Südekum, J. and Wößner, N. (2017). "German robots – The impact of industrial robots on workers". Technical report. Nuremberg: Institute for Employment Research.

[26] Bloom, N., Garicano, L., Sadun, R. and Van Reenen, J. (2014). "The distinct effects of information technology and communication technology on firm organization". Management Science, 60(12), 2859-2885.

[27] Trantopoulos, K., Krogh, G. von, Wallin, M.W. and Woerther, M. (2017). "External knowledge and information technology: implications for process innovation performance". MIS Quarterly, 41(1), 287-300.

[28] Ter Weel, B. (2015). "De match tussen mens en machine". Beleid en Maatschappij, 42(2), 156-170.

[29] Kuipers, H., Van Amelsvoort, P. and Kramer, E.-H. (2018). "Het Nieuwe Organiseren: alternatieven voor de bureaucratie". (3rd. Ed.) Leuven: Acco.

[30] Maenen, S. (2018). "Van Babel tot Ontwerp. Concepten en methoden voor organisatieontwikkeling". Kalmthout: Pelckmans Pro.

[31] Oeij, P.R.A., Rus, D. and Pot, F.D. (eds) (2017). "Workplace Innovation: Theory, Research and Practice". Cham (Switzerland): Springer.

[32] Porter, M. (1985). "Competitive Advantage". Boston, MA: Free Press.

[33] Vergeer, R., Kraan, K.O., Dhondt, S. and Kleinknecht, A. (2015). "Flexible labour and a firm's labour productivity growth: The importance of the innovation regime". European Journal of Economics and Economic Policies, 12(3), 300-317.

[34] Meijers, H.H.M. (1994). "On the diffusion of technologies in a vintage framework: theoretical considerations and empirical results". Maastricht: Datawyse / Universitaire Pers Maastricht.

[35] Belu, C. (2015). "Are distance measures effective at measuring efficiency? DEA meets the vintage model". Journal of Productivity Analysis, 43(3), 237-248.

[36] TNS Infratest (2014). "The IAB Establishment Panel. Employment Trends". Munich: Employer Survey.

[37] Nilutpal, D., Falaris, E.M. and Mulligan, J.G. (2009). "Vintage effects and the diffusion of time-saving technological innovations". The B.E. Journal of Economic Analysis & Policy, 9(1), 1-37.

[38] CBS (2018). "Internationaliseringsmonitor". Voorburg: Statistics Netherlands (CBS).

[39] Corrado, C., Haskel, J., Jona-Lasinio, C. and Iommi, M. (2012). "Intangible capital and growth". In Advanced Economies: Measurement Methods and Comparative Results. Discussion Paper No. 6733, IZA, Bonn.

[40] OECD (2015). "OECD Science, Technology and Industry Scoreboard 2015: Innovation for growth and society". Paris: OECD Publishing.

[41] Gómez, J. and Vargas, P. (2012). "Intangible resources and technology adoption in manufacturing firms". Research Policy, 41(9), 1607-1619.

[42] OECD (2013a). "Supporting Investment in Knowledge Capital, Investment and Innovation". Paris: OECD Publishing.

[43] Elbourne, A. and Grabska, K. (2016). "Evidence on Dutch macroeconomic and sectoral productivity performance: some stylised facts". Background document, CPB Netherlands Bureau for Economic Policy Analysis, Den Haag.

[44] Andrews, D. and Westmore, B. (2014). "Managerial capital and business R&D as enablers of productivity convergence". OECD Economics Department Working Papers, No. 1137, OECD Publishing, Paris.

[45] Saia, A., Andrews, D. and Albrizio, S. (2015). "Public policy and spillovers from the global productivity frontier: Industry level evidence". Working Paper No. 1238, OECD Economics Department, Paris.

[46] Byrne, D. and Corrado, C. (2016). "ICT prices and ICT services: What do they tell us about productivity and technology?" Economics Program Working Paper Series (Revised July 2016), The Conference Board, New York.

[47] Dhondt, S., and Kraan, K.O. (2019). "Gids monitoren technologisering en arbeidsmarkt". Leiden: TNO.

[48] Van Emmerik, M.L., Vroome, E.M.M. de, Kraan, K.O. and Bossche, S.N.J. van den (2017). "Werkgevers Enquête Arbeid 2016; Methodologie en Beschrijvende Resultaten". Leiden: TNO.

[49] Aasen, T.M., Møller, K. and Eriksson, A.F. (2013). "Nordiske strategier for medarbeiderdrevet innovasjon – 2013". København: Nordisk ministerråd, 60 p.

[50] Ter Weel, B., Van der Horst, A. and Gelauff, G. (2010). "The Netherlands of 2040: Thinking Ahead in the Netherlands". Den Haag: CPB Netherlands Bureau for Economic Policy Analysis.

[51] Child, J. (1972). "Organizational structure, environment and performance: the role of strategic choice". Sociology, 6(1), 1-22.

[52] Huys, R. (2001). "Uit de band? De structuur van arbeidsverdeling in de Belgische autoassemblagebedrijven". Leuven: Acco.

[53] Borghans, L. and Ter Weel, B. (2006). "The division of labour, worker organisation, and technological change". The Economic Journal, 116(509), 45-72.

[54] Felstead, A., Gallie, D., Green, F. and Zhou, Y. (2007). "Skills At Work, 1986 to 2006". Oxford: University of Oxford, SKOPE.

[55] OECD (2013b). "Technical Report of the Survey of Adult Skills (PIAAC)". Paris: OECD Publishing.

[56] Oudejans, M. (2012). "Value and Effectiveness of Work Tasks. Vragenlijst afgenomen in het LISS panel". Tilburg: Centerdata.

[57] Kleruj, N. (2017). "Nederlandse Skill Survey (NSS) 2017. Vragenlijst afgenomen in het LISS panel". Tilburg: Centerdata.

[58] Lorenz, E. and Valeyre, A. (2005). "Organisational innovation, human resource management and labour market structure: a comparison of the EU-15". The Journal of Industrial Relations, 47(4), 424-442.

[59] Dhondt, S., Kraan, K.O. and Preenen, P.T.Y. (2019). "Rapport Monitor Technologisering en Arbeidsmarkt". Report R678910 for Institute GAK. TNO, Leiden.

[60] Ter Weel, B. and Kok, S. (2013). "De Nederlandse arbeidsmarkt in taken: Eerste bevindingen uit de Nederlandse skills survey" *[The Dutch labour market in tasks: Preliminary findings from the Netherlands skills survey]*. Den Hague, The Netherlands: Netherlands Bureau for Economic Policy Analysis.

13

Digitalization and Management of Innovation: The Role of Technology, Environment, and Governance

Hans Schaffers

Adventure Research, The Netherlands

Abstract

The focus of this chapter is on the organization of innovation in an increasingly digitalized and networked world. Products, services, business processes, workplaces, and supply chains become connected, enabled by digital technologies and platforms, and leading to new ways of value creation based on new types of business models. Digitalization also affects the public sector, e.g., in domains such as health care, energy, education and transport, and urban life in the "smart city." Digitalization reshapes the process of innovation through facilitating new organizational forms such as collaborative networks and platform-based ecosystems. This chapter discusses examples of the new forms of managing innovation management in industry, the public sector, and also addresses the relatively new form of blockchain innovation. The aim is to understand the forces at work affecting how digital innovation in collaborative settings is organized to clarify the implications for managing innovation and to provide future-oriented recommendations to innovation stakeholders.

Keywords: Digitalization, innovation, management, networks and ecosystems, collaboration, governance

13.1 Introduction

Innovation is a societal, actor-driven phenomenon which can take different forms according to how it is shaped by advances in science and technology, entrepreneurial strategies and societal challenges, or combinations. It is widely recognized that innovation as an activity goes far beyond research and development (R&D), as innovation also covers adoption, implementation, and value capturing [1, 2]. Innovation activities in many cases go beyond the borders of single organizations, sectors and institutions, and it requires the integration of different technologies and knowledge areas. The object or "artifact" of innovation comprises not only technologies but includes also—incrementally improved or radically renewed—architectures and components, products and services, processes and practices, infrastructures, work organizations and production systems, and business models. Increasingly, innovation is situated in networks and ecosystems initiated and orchestrated by large companies, enabling them to benefit from collaboration opportunities in the creation of knowledge, innovations, and eventually value in the market [3]. Small- and medium-sized (SME) companies contribute to and benefit from the new innovation environments by adapting their knowledge sourcing and innovation processes [4].

With the changing innovation culture and practices over the last decades, innovation systems and innovation policies thinking have significantly shifted emphasis from the concept of systems of innovation [5, 6] toward the paradigm of open innovation [3]. Given the forces of globalization and increasing complexity of societal challenges, policy focus has been shifting from the national level increasingly to sectoral policies and to regional and supranational contexts. Modern innovation thinking strongly emphasizes the early involvement of users and diffusion of innovations in networks (e.g., [7]). The increasing tendency toward openness of innovation networks has also promoted interest in emergent and self-organized network governance, making it difficult to predict their evolution. Within this context of open, networked innovation, the nature of innovation policy instruments as well as the role of the state versus other actors within ecosystems of innovation has met a lot of attention. The challenging work of Mazzucato [8] demonstrates the key role of public funding of research and innovation by risk-taking "entrepreneurial states" in promoting technological innovation, and the key role of innovation ecosystems building upon symbiotic rather than parasitic public–private partnerships to sustain the impact over the longer term.

To cope with the complex and systemic nature of innovations, the concept of "systemic instruments" has been proposed (e.g., [9]). It addresses the structure and function of policy instruments, in particular the organization of innovation systems, the management of its interfaces across subsystems, the capability for visioning, learning and experimenting, and the demand orientation. This type of work is useful as a conceptual framework for analyzing systemic problems in innovation. At the same time, there remains a need to better understand the role of decision-making, planning and common visioning mechanisms in practice, as well as the role of cross-organizational culture differences, which affect the evolution of multiactor innovation ecosystems.

In this context, digitalization and digital innovation bring new challenges. While digital technologies affect products, services, and processes, these technologies also drive platform ecosystems and transform organizations and business models, as well as reshape the process of digital innovation itself. Many companies experience difficulties in benefiting from the changed nature of innovations as digital, systemic, and network oriented. Products, services, and processes are becoming connected and smart through the integration of digital technologies [10, 11] and those innovations can be termed "digital innovations." Such digital innovations as objects or artifacts are not confined to stand-alone product, process, and service innovations but form part of larger systems and include business model innovations in a context of digitally supported supply chains and value networks. The economic, societal, and technological environment of digital innovation has brought platforms and ecosystems to the foreground, and how strategies for digital value extraction are building on network effects and are considering users as products and user data as assets. This way a small number of companies such as Facebook and Google have achieved monopoly positions, as is well explained by among others Mazzucato [12] and Zuboff [13]. Such developments are not only relevant in the business and economic domains but are, given the antidemocratic implications, societally and politically as well. To understand these changes and the forces at work and how they affect the process of innovation in an increasingly digitalized world, this chapter discusses how "digital innovation" is organized within a changing economic and societal environment. Three areas of digital innovation provide examples: industrial innovation in the context of Industry 4.0, urban innovation aiming the realization of "smart cities," and blockchain applications constituting decentralized (or even sovereign) forms of organizing.

13.2 Digital Innovation as Multidimensional Concept

13.2.1 Digital Technologies and Digital Innovation

The term "digital technologies" comprises a broad spectrum, including intelligent data sensing and processing, electronic components and systems, computing and software, networking and security, and blockchain. Digital technologies give shape to "innovation" in multiple ways as it affects the nature, but also the process and organization of innovation. It affects the nature of innovations in terms of objects or artifacts. Most visibly in our daily lives are "smart" devices based on the concept of Internet of Things, such as smart mobiles, thermostats, door locks, watches, cameras, and more. In industry, digital technologies enable machine-to-machine communication and connection between physical resources, business processes, and factories [11]. With digital technologies embedded, products or systems are becoming modular, compatible, interoperable, and connected. This forms the basis for servitization-based business models where use of services rather than owning products delivers value.

Digital technologies underlie a broad diversity of devices, applications, and services, which enables the myriad of monitoring, communication, interaction, and transaction processes of high importance for business, consumers and citizens, governments, and for society at large. Digital technologies thus lie at the basis of transformational innovations, processes, organizational structures, and business models in industry but also in sectors such as energy supply, transport and logistics, and health care. Examples include smart systems for distributed energy management, traffic and mobility control systems in cities, advanced robotics and machines as part of smart factories, and digital platforms and marketplaces enabling demand-driven supply chains. These applications and processes are based on the ability for sensing and processing of and acting upon large amounts of data and the ability to connect smart objects of all kinds into what is called the Internet of Things, Services, and People [14].

13.2.2 Digital Technologies and the Innovation Process

The advance of digital technologies does not only affect the object of innovation but also transforms the innovation processes and networks of innovation. A prominent development is the rise of platforms and platform ecosystems [15, 16]. Digital platforms exploit the mechanism of multisided markets and network externalities to bring together different types of actors such as service providers, product developers, customers and technology innovators, with complex interorganizational relationships orchestrated in such a way as to benefiting the platform ecosystem. As products become smart, interconnected, and platform-based, they generate a range of new opportunities for innovative product-service systems, which also contribute to changes in business processes. This development is not only making processes more efficient and interconnected, but also laying the foundations for networked organizations and agile, demand driven supply chains. Supply chains and value networks become digitalized, platform-based, demand driven, and configurable. As the costs of communication and value of remote information decrease, digital technologies increasingly enable distributed, networked forms of organizing [17]. Essentially, these changes are not only industrial but also social and societal. Organizational and social innovation such as in new ways of collaboration and collaborative working or new ways of "peer production" as analyzed by Benkler [18] gets more to the foreground compared to technological and product innovation.

The transformation of innovation processes and networks is potentially providing opportunities for a broader range of stakeholders to engage in innovation practices and making these practices more inclusive, cocreative, and responsible [1]. Over the last couple of decades, the innovation process has already undergone a transition from closed and linear technology push toward more user driven, collaborative, and iterative. Innovation is governed by increasingly shorter time horizons and need for fast adaptations, enabled by flexible forms of cooperation in often temporary partnerships, and supported by web-based tools and data analytics. Increasingly, innovation is informed by societal challenges. This is bringing the need for large-scale transitions, e.g., in domains of health care, energy, and climate change. Furthermore, ethical issues surrounding the development and application of technologies such as artificial intelligence and big data demands for new forms of responsible innovation. Due to this changing landscape the idea of open, collaborative innovation ecosystems has come to the foreground. The continuous drive toward innovation has become a key enabler of renewal in modern ways of working, organizing, and production. Innovation has become a much more open and collaborative process which crosses the boundaries of internal R&D departments. Accelerated by the introduction of digital technologies, the situational context of innovation has changed from the company toward supply chains, business networks, and nowadays digitally supported platform ecosystems.

In a broad sense, the term "digitalization" captures the transformation toward an economy and society enabled by digital technologies and infrastructures which facilitate the network paradigm [19]. As a social process, the innovation process is part of this digital transformation. Digitalization has changed the process of innovation itself due to its impact on (connected and interoperable) products, services and processes, but also due to the way interaction, communication, and creativity among people in teams across organizations and in communities is enabled. For example, digital platforms facilitate innovation marketplaces and crowdsourcing in new ways. The combined impacts of digitalization enable a range of new forms of distributed innovation governance such as platform-based, regionally based, and globally connected innovation ecosystems, targeting both radical and incremental innovations and fundamental as well as application-oriented levels.

The interplay of different factors mentioned has stimulated the emergence of alternative forms of innovation governance to emerge such as the network, the platform ecosystem, and the crowdsourcing community. New principles of organization emerged around the concept of collaborative innovation and value creation. New policy approaches have been implemented to address societally oriented systemic forms of innovation and ecosystem building. All this has influenced the demand driven, collaborative, and open forms of innovation becoming more and more popular. The changing environment and scope of innovation invites us to identify and understand the new practices of innovation. A deeper understanding of evolving practices will also provide us with new ways of organizing innovation and policy making suitable to the context of digitalization.

13.3 Theoretical Viewpoints on Managing Digital Innovation

13.3.1 Conceptual Analysis of Digital Innovation

During the past two decades, digital innovation increasingly has become a topic of scientific debate. Still, there have been mostly conceptually oriented studies of managing digital innovation, aiming at "making sense" of the phenomenon, while empirical studies have been relatively scarce. For the purpose of this chapter, our focus is on the interdependencies between digital innovation, the innovation environment, and the governance of innovation. The actual digital innovation process is shaped by these interdependencies and shapes them as well. To illustrate these interdependencies a selection of publications is categorized in Table 13.1 which will be concisely discussed.

Several authors have proposed conceptual definitions to better capture and understand digital innovation as object or artifact, and as process. Based on how digitization affects product architecture in terms of a layered modular architecture made up of devices, networks, services, and content, an influential conceptualization of digital innovation was developed by Yoo et al. [20]. Their focus on product innovation and architecture leads them to define digital innovation as "the carrying out of new combinations of digital and physical components to produce novel products." The layered modular architecture enables digital product platforms, which form the basis of business ecosystems. Equally, Nambisan et al. [22] interpret digital innovation as both the use of digital tools and infrastructure during the process of innovating (such as 3D printing, data analytics, and mobile computing) and the outcome of innovation. Digital innovation in the sense of outcome is understood in a broad sense as the creation of market offerings, business processes, platforms, or business models that result from the use of digital technologies.

The implication is that a discussion of digital innovation inescapably will be intertwined with discussing platforms and ecosystems as an environment for digital innovation. As explained by Yoo et al. [20], digital innovation is distributed across firms of different kinds in a network or ecosystem. This line of thinking was continued in further focusing on the role of digital platforms in shaping ecosystems [21]. These ecosystems attract heterogeneous actors and offer digital tools, allowing firms to create platforms of not just products but of digital capabilities as well. One key aspect of how digital technologies enable innovations is characterized by "convergence," which captures the ability to combine separate components within one system (such as a smartphone). Another aspect is "generativity," which identifies the (software-enabled) extensibility and adaptivity of those innovations. All this implies that digital innovation situated in platform environments follows a different organizing logic compared with traditional stage-gate models.

13.3.2 Innovation in Networks and Ecosystems

The role of networks and ecosystems as organizing principles for value creation has been studied intensively. For example, Adner and Kapoor [35] investigated how the success of innovators is in many cases depending

Table 13.1   Selection of publications on managing and organizing digital innovation.

Themes	Authors	Main aspects addressed
Digital technologies and the nature of digital innovation	Yoo, Henfridsson, and Lyytinen [20]	Digital technology instigates a new layered modular type of product architecture. This implies fluidity in digital innovations and driving decentral, distributed forms of innovation.
	Yoo et al. [21]	Digital platforms offering tools and capabilities to a variety of innovation actors
	Nambisan et al. [22] and Nambisan et al. [23]	Digital tools and infrastructures and orchestration capabilities enable the process of digital innovation.
Innovation in networks and ecosystems	Autio and Thomas [24] and Thomas and Autio [25]	Characteristics of innovation ecosystems and implications for management of innovation
	Macedo and Camarinha-Matos [26]	Building and maintaining of trust, incentives, rules of fairness, and sharing benefits, transparent governance principles, and alignment of values of participants and collaborative networks
	Gawer and Cusumano [15], Tiwana [16], andIschkia et al. [27]	Platform architecture and governance mechanisms enabling innovation on complementary products and services
Digital innovation management	Nambisan et al. [22]	Digital innovation management research should address the fluid nature of digital innovation, the distributed nature of innovation agency, and the heterogeneity of actors involved.
	Yoo et al. [21]	Characteristics of digital innovation lead to increasing importance of digital platforms, distributed forms of innovation, and prevalence of combinatorial innovation.
	Nylén and Holmström [28]	Digital technologies enable new types of innovation processes; therefore, firms need new skills and tools to support them.
	Svahn et al. [29]	Digital innovation requires the balancing of new and established capabilities and core competencies, balancing the process and product focus, and balancing external and internal collaboration.
	Kane et al. [30]	Digital ecosystems and cross-functional teams are key sources of digital innovation. But while digitally mature companies are more agile and innovative, they require greater governance.
Digital innovation and the workplace	Arundel et al. [31], Oeij et al. [32], Kopp et al. [33], andPot et al. [34]	Work organization may stimulate innovation. Digitalization and digital transformation require workplace innovation.

on the efforts of other innovators in its environment and the relation between them, including technological interdependence and distribution of value. An insightful study of the role of network resources in shaping technological innovation systems is presented by Musiolik et al. [36] focusing on energy innovation. They find that in different types of networks the organizational resources of network members and network-level resources such as governance mechanisms, trust building, and common vision development play a role in shaping innovation systems in different ways.

Of high interest for managing digital innovation is how governance in networks should be organized as to create effectiveness of the network in terms of strength of collaboration among its members. According to Provan and Kenis [37], the network is considered as highly suited to resolving "wicked" and complex problems, e.g., in the context of product innovation. Different principles may guide network governance: self-regulation by all participants, management by a lead organization, or by an independent agency, whose impact on network effectiveness depends on the situational context. The effectiveness of specific governance forms may depend on trust requirements, number of actors involved, consensus level required, and need for network competencies. Studies on collaborative networked organizations addressed the life cycle of creating and management of such networked organizations, and the underlying conditions such as trust and collaborative culture, which could be applicable to innovation ecosystems as well [26, 38]. Although there is some debate about its usefulness [39], the concept of innovation ecosystem has emerged as a useful description of the current environment of competition and innovation [24, 25].

Until now, the issue of how innovation ecosystems raise implications for management of innovation has been addressed only in limited ways. Autio and Thomas [24] define an innovation ecosystem as "a network of interconnected organizations, organized around a focal firm or a platform, and incorporating both production and use side participants, and focusing on the development of new value through innovation." It is, however, concluded that "too little is still known about how firms can proactively create, steer, and leverage innovation ecosystems for enhanced innovation performance" and there seems to be a gap in understanding the creation and management of innovation ecosystems, the process of value appropriation, and the role of control mechanisms such as platform and critical assets control in appropriating value.

Studies explicitly related to digital innovation in networks and ecosystems are still scarce. A reason might be that platform-based ecosystems, which are dominated by single enterprises have emerged as a dominant innovation ecosystem for digital innovation. A number of publications have discussed the management of innovation in relation to platform ecosystems. Building on the concept of multisided markets developed by Eisenmann et al. [40], platform literature emphasizes the mutual interest in value creation by the platform organizer, external contributors, and customers. For example, Gawer and Cusumano [15] discuss how platform leaders develop platform vision, build technical architecture and interfaces, create a platform coalition, and thus evolve the platform. This way industry platforms facilitate and increase the degree of innovation on complementary products and services, and function as a technological foundation at the heart of innovative business ecosystems.

13.3.3 Managing Digital Innovation in Ecosystems

Given the role of platforms and ecosystems in digital innovation, it can be expected that the process of innovation will differ from the model proposed by Tidd and Bessant [2]. This model structures the innovation process in the four phases: search, select, implement, and capture. However, in a platform environment we have interorganizational relationships determined by, on the one hand, the complementary positions and relationships of platform actors, and, on the other, by the platform architecture and governance mechanisms [16]. Platform ecosystems are considered as evolving organisms that cannot be purpose built and rigidly managed, but rather considered as sets of resources that can be orchestrated toward a common goal. This represents a much more fluid environment for innovation than indicated by the mentioned phase model.

Digital innovation management in terms of orchestration practices and processes with respect to digital innovation is conceptually elaborated by Nambisan et al. [22]. The authors aim at connecting information systems theory with organizational theories in explaining how digital tools enable or constrain innovation outcomes, and how the notion of orchestration sheds light on distributed forms of digital innovation. Their research agenda proposes a deeper understanding of sense making, decision-making, and problem solving in digital innovation processes and in orchestration of innovation management.

Empirical research on digital innovation process and management is scarce. Kane et al. [30] found that that digitally mature companies stimulate innovation throughout the company, not just in R&D departments, and in many cases in cross-functional teams and through collaboration with external parties. Also, such companies are innovating at a far higher rate and more agile than less digitally mature companies. Actually, collaboration with external organizations and within ecosystems is found to be a key source of digital innovation.

Several studies have been devoted to various aspects of managing digital innovation. Isckia et al. [27] analyses how platform-based ecosystems can ensure the commercialization of a flow of digital innovations based on how platform-owners orchestrate the ecosystem in terms of enabling interactions among different autonomous players, including individuals, communities, and organizations. Technically this is enabled by the platform architecture, which offers modularity as coordinating engine of partners, as well as generativity. Pellika and Ali-Vehmas [41] investigate how the ability to successfully innovate and commercialize a product not only depends on the firm's technological strategy and effective innovation process but also on its capabilities to manage the innovation ecosystem in the style of a quadruple helix. In their view, this requires collaborative structures and processes adhered to by the organizations that are part of the ecosystem, to agree on partner roles, coordinate resources, and partner investments, and align timing of market entry. They conclude that innovation ecosystem strategy requires emphasis on common vision building, and capabilities to manage assets and resources outside its direct control. Cocreation, networking, and interaction with ecosystem partners are crucial. Valkokari et al. [42] address the ecosystem competences and ecosystem positioning required by the actors in the ecosystem to manage dynamic strategic interactions related to innovation. The authors' view is that ecosystem orchestration implies sense making of the innovation ecosystem by all actors and willingness not only focusing on their own business but to jointly develop the ability to design and manage innovation communities. Given the nature of digital innovation within a context of ecosystems, it appears relevant to study how ecosystem relations can be influenced to stimulate open innovation. Selander et al. [43] describe a case study looking at the process by which firms transform ecosystem relationships. Their conclusion is that such transformations require the resolution of value tensions among ecosystem participants.

Digital strategy has become a key topic in literature, focusing mostly on sources of value creation and business models in digital strategy. However, the connection with digital innovation is often neglected, and the issue how digital product and service innovation could be part of digital strategy development is a topic mostly discussed in conceptual terms. For example, Nylén and Holmström [28] stress the complexity of digital innovation in terms of the rapid pace of it, stressing the ease of reconfiguration and generativity of digital technology, which leads to cascades of innovations. This is also analyzed by Nambisan et al. [21]. A digital innovation strategy could be based on user experience, value proposition, digital evolution scanning, skills, and improvisation. Svahn et al. [29] understand digital innovation as an organizational capability, not merely a technology platform or an innovation incubator, and present an interesting case study at Volvo Cars focusing on how to balance a new set of innovation challenges: development of new versus already established innovation capabilities, focus on internal versus external collaboration, and pursuing flexibility versus control strategies. The misalignment of organizational capabilities to respond to digital opportunities is discussed by Kohli and Melville [44]. One of the points raised is how knowledge sharing in communities may enhance digital innovation. Their study concludes that there are knowledge gaps in digital innovation including the initiation and exploitation of digital innovation.

13.3.4 Digital Innovation in Collaborative Working Environments

While the previous section focused on the wider innovation environment, innovation as a human collaborative activity is enabled by the working environment. Due to automation of work, digitization of processes, and coordination by platforms, work and employment are affected in terms of tasks and working conditions [45]. Clearly, platform-based crowd working may negatively impact on the quality of work and working conditions and they also are not that relevant for innovation. On the other hand, digitization of processes opens them up to interesting forms of collaborative decentralized production, e.g., the "makers" movement and hacker communities [45]. In this respect, the impact of digitalization on shaping new forms of collaboration and production can be expected to create new opportunities for decentralized innovation. In particular, the commons-based peer to peer production model, which is highly adept to learning and innovation in rapidly changing, uncertain, and complex environments, as promoted by Benkler [18] and Bauwens et al. [46], is highly relevant.

Earlier work based on aggregate indicators in a study of EU-15 countries already made clear how organizing affects innovation behavior [31]. It appears that in nations where work is organized around learning on the job and independently solving complex problems firms tend to be more active in terms of endogenous (in-house) innovation. The implication is that improving the innovative capabilities of European firms might lie in new forms of work organization and employee participation that stimulate innovation. Comparable conclusions in the context of workplace innovation and digital innovation have been reached by Cox et al. [47], Oeij et al. [32], Kopp et al. [33], and Pot et al. [34]. The common message is that digital innovation and digital transformation requires workplace innovation. Emphasis in these studies is mostly on creating the workplace conditions that underly digital transformation, less on the actual process of innovation and how it is intertwined with working and learning in the workplace. In the next sections, the evolving practices in digital innovation management will be discussed based on examples taken from the domains of industry innovation, smart cities innovation, and blockchain innovation.

13.4 Digitalization and Managing Industrial Innovation

13.4.1 Industrial Innovation in Ecosystem Environments

Traditional industrial forms of innovation have been centered around the R&D function within companies, focusing on product and process innovation. In the closed R&D lab, common goals, values, and beliefs ensure trust and create effective collaboration. As soon as R&D becomes part of wider interactions within or outside the organization, different incentives, rules, and procedures become necessary. Traditional R&D management mostly dealt with the development, selection, and management of R&D projects. The dominant management approach in this context has been the stage-gate or "waterfall" model. This approach matches the traditional governance model of hierarchical organizations that is still widely accepted. In line with this, Tidd and Bessant [2] present a view of the innovation process as turning ideas into reality and capturing value, phased in four steps: search, selection, implementation, and value capturing. The mission of innovation management is to offer the practices, routines, and tools to influence the innovation process.

Clearly the innovation process has transformed toward open, collaborative, networked, agile, and iterative forms in a context of networks, platforms, and ecosystems. Open innovation, a term coined by Henry Chesbrough [3] to indicate the increasing external cooperation in the innovation process, has become a key trend transforming traditional forms of inward oriented industrial R&D. Enterprises have become much more sensible to innovation opportunities, e.g., by engaging employees and stimulating workplace innovation, and by anticipating on rapidly changing customer demands.

The processes of research, development, and innovation (including adoption and implementation) have also become more intertwined with business operations including production and marketing. Innovation has become a continuous and cyclic experimentation and testing of improvements and radical changes carried out in multidisciplinary teams across boundaries of organizations and in collaboration with users. Over time, the open innovation concept has further evolved toward distributed forms of innovation based on cooperation among innovation partners in "innovation ecosystems" and platform-based forms such as crowdsourcing, highly enabled by digital tools and infrastructures. As a result, stakeholder and community-oriented collaboration settings for innovation are more common, such as innovation platforms, crowdsourcing, innovation competitions, open test beds, living labs, field labs, maker spaces, and learning factories. Innovation has more commonly become an activity situated in innovation communities, networks, and platforms.

This raises the issue how traditional industry sectors could transform toward ecosystem orientations and which hurdles are to be overcome and how. One of the projects, which empirically studied the changes in industrial innovation, has been the Industrial Innovation in Transition[1] (IIT). The project conducted more than 700 interviews and generated useful empirical information based on data from 5 sectors and 11 European countries. Some of the most interesting findings are the following [48]:

- Ecosystem relations have become more important for innovation for companies in all sectors. Companies are making use of their innovation ecosystem, for anticipating their future environment and for benefiting from the knowledge flows from and to suppliers and customers.
- They also shape their ecosystem, e.g., by common vision building, network building, and business models oriented to ecosystems.
- It appears that innovative companies are part of a broader ecosystem, where key partners include value chain partners such as customers and suppliers, but also public research bodies, investors, regulators, and policy makers.
- The majority of companies indicated that they are already engaging in open innovation activities, e.g., through increasing reliance on

---

[1]The chapter author has been a codeveloper of the Industrial Innovation in Transition project (Grant Agreement No. 649351) and has been involved as a researcher during its first phase.

knowledge from outside the company. But not that much for developing core technologies; rather for solving technical problems and expanding market prospects. Reasons for that reside in the willingness to remain in control over critical technologies. Other issues are how to distribute costs, benefits and risks, and how to acquire the capabilities and competences needed for open innovation.

- New models and tools for innovation management include the adoption of cocreation principles, the application of web-enabled tools to support innovation communities, and the increasing use of big data.
- Regarding the organization of innovation management, it appears that the stage-gate model is still used extensively, but also the customer-driven lean start-up model. Main decision-making actors are innovation committees and R&D departments.

These findings still seem to demonstrate company-centric open innovation strategies although there are evidences pointing to a growing ecosystem orientation. The use of digital technologies to support collaborative innovation has been growing as well. The IIT project did not explicitly focus on digital innovation, however, it can be expected that the ongoing efficiency-driven digitalization of supply chains could further stimulate digital product, service, and process innovations.

13.4.2 Challenges to Create Industry 4.0 Business Ecosystems

The current development toward Industry 4.0 is mostly driven by innovative manufacturing technologies and ICTs, for which the Internet of Things concept is the key driver. Industry 4.0 emphasizes how smart factories are connected within a wider business ecosystem to enable network-centric production methods, smart product and service concepts, and responsive business models in collaborative ecosystems [49]. While Industry 4.0 scenarios still lie largely in the future, it is expected a further trend toward factories become part of a smart, network-centric production environments, enabling business models at the level of platforms and ecosystems.

One of the more visible manifestations of Industry 4.0 scenarios in practice is "servitization," the transformation of product-based toward service- and solution-based business models. The implication is that supply chains become more connected and ecosystem oriented and must build upon flexible forms of cooperation across firm boundaries [50]. This raises many challenges. Underlying infrastructures and shared applications of the ecosystem participants must be interoperable to enable sharing of data, and organizational arrangements and processes must be synchronized to enable collaboration, and sharing of competencies, roles, and responsibilities in the ecosystem must be agreed [51]. Additionally, transition to servitization requires practical agreements regarding distribution of risk, costs, and benefits among collaborators, and agreements concerning membership and exit of partners.

As discussed, digitalization and servitization have brought platform-oriented business models to the forefront, and it can be expected to also transform traditional industrial supply chains. Platform companies have now become powerful and even dominant economic entities. Platform ecosystems have become a new type of business environment but its implications for innovation and management of innovation are still not too clear. Governance, regulation, and orchestration of platform ecosystems will probably play a key role in stimulating industrial innovation. Although there is only scarce empirical evidence so far, it can be expected that platform business models imply a more intensive level of innovation cooperation across the ecosystem. As platforms bring together different groups of technology providers, sellers, external innovators and customers, and embody network effects, several issues must be agreed on among partners not just for doing business but also for management of innovation: openness, modularity, control, sharing, and interoperability [16]. Platforms thus lead to rethink innovation management in terms of organizational structures, leadership, business models, and value capture.

It should also be mentioned that platforms besides those based on digital infrastructure can take more traditional forms of collaboration. For example, in Germany a new model of cooperative social innovation has emerged called learning factories. The learning factory concept as practiced in Germany aims at local learning, education, and innovation in manufacturing based on strong collaborations among professional schools, businesses, and SMEs [52]. It focuses on bringing learning, innovation, education, and training closer to industrial practice in realistic manufacturing environments. In the Netherlands, Smart Industry Field labs have been established as environments in which companies and knowledge institutions jointly innovate. Field labs aim at achieving breakthrough innovations, strengthen innovation ecosystems, connect professional education and business, and increase human capital development. Around 40 field labs are existing in areas such as personalized products and services, dairy farming, secure connected systems, metal 3D printing, predictive maintenance, and connected supplier networks.

13.5 Digital Innovation in Urban Environments

13.5.1 Cities as Complex Systems

Nowadays it is common to consider cities as complex systems, grounded in large social networks and decentralized networked infrastructures [53]. These systems are continuously changing under the influence of political, social, demographic, economic, and technological forces. The transformative power of urbanization has in part been facilitated by the rapid deployment of information and communication technologies [54]. Such technologies may help cities addressing a diverse range of complex challenges, such as sustainable and circular economy, energy distribution, water management, health care, housing, public transport and mobility, environment, and climate. Digital technologies based on sensors, wireless networks, smart devices, and data analytics add the capability to predict and control the functioning of such systems in complex urban networks, and predict demand, need for maintenance, repair, and replacement. In this respect, the Internet of Things is becoming a major driver of data-driven digital innovation in cities, as it integrates diverse technologies, systems, and functionalities with human behavior in addressing urban challenges. Cassandras [55] even proposes the Internet of Things perspective of smart cities as cyber-physical systems, of which the infrastructure is based on a network of sensors and actuators interacting with smart devices and integrated with cloud service infrastructure. A major issue for this paradigm's feasibility is not only the need to create a common software environment, but also its social acceptation, given largely unsolved privacy and data ownership concerns.

Ongoing critical discussions of the concept of "smart city" stressed its technology and business-driven perspective [56]. Based on ideal futuristic blueprints, architect Rem Koolhaas made the observation that "the citizens the smart city claims to serve are treated like infants" [57]. In his view, the city looks like a comprehensive surveillance system as smart cities and politics seem to have been growing in separate worlds. The dominance of data-driven platform companies and the increasing awareness of control, privacy, and dominance issues are reflected in the concept of "surveillance capitalism" as proposed by Shoshana Zuboff to describe a new form of capitalism based on the capabilities to predict and control human behavior [13]. Indeed, many smart city initiatives have been driven by the ambition to demonstrate the applicability of technologies such as in the scope of Internet of Things in urban areas, and not necessarily by the need to address urban challenges and improve urban living.

In the last decade there has been growing an increasing awareness of the importance of engaging local communities and stakeholders in urban innovation. Availability of easy-to-use software tools is empowering these communities and stakeholders to collaborate, share ideas and information and participate in urban planning and innovation activities, and to establish local community networks. This has definitely contributed to the possibility of more "democratic" urban innovation environments. Whereas the first phase of this development started with technology-driven "smart city" scenarios, the second phase witnessed the testing of technologies and applications in small scale city pilots. The third phase could be described as is the emergence of—mostly still small-scale—urban living labs for participative innovation, often enabled by those easy-to-use developer tools. A fourth phase may bring forms of self-organized "democratic" innovation based on engagement in local community networks.

13.5.2 Organizing Digital Innovation in Urban Environments

Technologies such as Internet of Things, wireless and optical networks, and data analytics are helpful to optimize and control infrastructural urban systems. A step further, the idea of "city as test beds" [58, 59] has value to validate digital technologies in real life, although active citizen participation is often very limited. Humans, besides cars, buildings, roads, and other assets, primarily act as "sensing objects" [60, 61]. European Framework programs played an important role to stimulate technological experimentation and innovation in a range of technology areas, including networking technologies, the Internet, big data, cloud computing, and cyber-physical systems [62]. Increasingly cities have been identified as real-life playgrounds to test technologies, networks, and applications. One of the early but well-developed projects with respect to Internet of Things in cities has been SmartSantander in Spain, which did not only experiment on technologies and architectures but also on applications and services for various use cases such as environmental monitoring, parking management and driver guidance, precision irrigation, augmented reality, and participatory sensing [58, 63, 64].

A next phase of creating test bed environments is the FED4FIRE initiative, a large-scale European federation of 23 facilities and test beds providing one-stop-shopping experimentation services as experimentation-as-a-service [65]. This initiative also supports urban test beds. The CityLab test bed established in the city of Antwerp, Belgium, part of the City of Things program [59] offers experimenters experimentation tools based on the FED4FIRE results and enables testing of Internet of Things network functionalities in real-life conditions. With the advance of Internet of Things initiatives at the European level and in many cities, crowdsourcing and crowd sensing as a means to collect inputs from end users have increasingly become spearheads in relation to the city as lab concept. An even more important development has been the development of Internet of Things platform ecosystems within the IoT-European Platforms Initiative (IoT-EPI). These projects not only developed innovative IoT platform technologies but also fostered technology adoption through community and business building activities. A significant step toward the creation of ecosystems of urban innovation was taken by the SynchroniCity project within Horizon 2020, which aimed at creating a global marketplace for IoT-enabled services, enabling cities and businesses to develop shared digital services to grow local economies [66]. The open source FIWARE software platform, a result of the Future Internet Public Private Partnership, is another example. These examples demonstrate a transformation of the role of test beds toward becoming an infrastructure for platform ecosystems. This transformation will require ecosystem building and governance approaches to become effective [67]. So far, there are few signs that this has been accomplished; thus, it remains a challenge for the near future.

The transformation will also require the integration of user driven innovation principles, as promoted by the living labs movement, which requires the cooperation among communities involved with different orientations such as resolving specific urban challenges, technology development and testing, and user-led innovation [68, 69]. Within a setting of urban ecosystems and given the availability of easy-to-use software tools for application development we may expect forms of self-organization.

Actually, the living labs concept, like comparable terms such as "city-as-a-lab," is only weakly defined, so it is no surprise that it has expanded into different directions. It should also be recognized that cities' roles and approaches to living labs innovation can be diverse. Leminen et al. [70] distinguish four different roles: urban service provider, neighborhood participator and co-organizer of bottom-up activities, catalyst and platform organizer, and as rapid experimenter through setting up trials. De Waal and Dignum [71] discuss the role of smart city visions that may contain hidden values in shaping these roles. We may conclude that these visions will never be able to capture all the ambitions, needs and demands of citizens, entrepreneurs and other stakeholders, which is also very natural, and it is the challenge of the democratic process to address that. Different approaches toward implementing living labs concepts in urban development based on some kind of platform ecosystem vision have been pursued. Pragmatical approaches based on a web-based project development environment where participants can initiate innovative projects seem to work out well, although it remains difficult to ensure scaling-up of urban digital innovations. Such projects are often aimed at stimulating entrepreneurship and creation of start-ups and decentralized models of production and service.

The biggest challenge in terms of urban innovation seems to be to create and nurture the ecosystems that drive innovation, project development and successful implementation as well as experimentation and learning. The concept of a smart city innovation ecosystem should be handled with care, as the term "innovation ecosystem" often remains a goal to be realized rather than accomplished fact. Some authors argue that the concept of innovation ecosystems is flawed in itself, as it implies a faulty analogy to "natural ecosystems" and does not add much to the concept of "regional innovation system" [39].

This brings up the issue of how cities are organizing their policies and collaboration approaches in relation to urban development and becoming smart cities. It seems that smart cities policies are making cities better performing through innovation, in particular by increasing the stock of knowledge [72]. A global review of smart cities strategies performed by Future Cities Catapult [73] presents a number of insightful observations of how Smart City strategies are being developed in selected cities. It concludes that there are considerable differences in approaches of cities toward strategy development and implementation. Sometimes cities have a formal smart city strategy but more often smart city approach is embedded in municipal policy in general. Sometimes emphasis is on the innovation ecosystem to drive smart city solutions (Helsinki) but emphasis also can lie on coordinating smart city projects (Berlin, Manchester).

Given the understanding of cities as complex systems, we should neither expect top-down approaches to be very successful nor should smart city visions be taken too seriously. Pragmatical approaches will build upon the building and nurturing the collaboration among different groups of stakeholders, development of innovation capabilities including making data and digital tools available, and creating a portfolio of innovative projects from which lessons can be learned rapidly.

13.6 Digital Innovation in Blockchain Organizations

A relatively new type of business ecosystem is extremely decentralized and based on blockchain technology, also called distributed ledger technology. Although a lot of innovation is going on, e.g., in financial services and payments, and use case experimentations and piloting in many other sectors, blockchain is still in its early stages of maturity. It is not clear whether it will survive except in very specific applications such as in the financial domain. Still it is an interesting scenario for studying the management of digital innovation, for which findings discussed in previous sections may not or not fully apply, and innovation studies in this context are still very scarce. A blockchain is based on the principle of cryptographic secured transactions forming a growing chain of chronologically linked records (blocks). Each block contains a cryptographic hash of the previous block, a timestamp, and transaction data. Blockchain systems are based on some form of transaction validation based on consensus mechanisms, ideally in a distributed manner such as for Bitcoin [74]. A key innovation is the smart contract which enables algorithmic-based transaction execution. The ultimate relevance of blockchain goes far beyond Bitcoin as it in principle enables peer-to-peer economies and radically decentralized forms of organizing and business models. Current applications include a range of sectors among which financial transactions, business process management, auditing, insurance, energy transactions, education, supply chain management, manufacturing, asset sharing, real estate, and government services [75]. Blockchain applications are often crossing the borders of sectors and particularly at the interfaces many innovative use cases can be identified.

Basically, three different types of blockchain systems can be distinguished, according to participation and consensus mechanism involved. The first type is the public "permissionless" blockchain, such as the Ethereum and Bitcoin, which is decentralized, open for participation by anyone, and operated by the blockchain community. The second type is private "permissioned" blockchain, which brings together a closed group of participants and is operated by a single entity under central, member-based ownership. The third type is the federated or consortium blockchain, which is operated by consortium participants and establishes cooperation among specific parties under consortium leadership. This is the form we see, e.g., in logistics and supply chain management.

Blockchain ecosystems usually consist of several types of key players [74]: platforms such as the Ethereum or Hyperledger, financial institutions such as banking and insurance companies, venture capitalists, and computing firms, apart from leading companies in sectoral domains such as logistics and supply chain management. Also, regulators, standardization organizations, and governments are involved.

Innovation as artifact and process in the context of blockchain systems is relevant in different ways. First, as indicated earlier, blockchain technology and applications are still immature and under development. Second, blockchain can support the transactions-oriented management of R&D and innovation, e.g., in terms of tracking and management and exchange of research rights, legal contracts, and intellectual property rights. An example is the data rights management over Ethereum blockchain network [76]. An interesting vision of how blockchain technology could act as an organizational capability to enable open innovation for sustainability transitions is presented by Narayan and Tidström [77]. On the other hand, other more bottom-up forms of organizing open innovation are possible, as envisaged by Pazaitis [78].

Third, blockchain networks embody specific governance procedures and processes in terms of participation, access, and decision rights. The governance model determines the distribution of roles and responsibilities among network participants and this is different in different types of blockchain platforms as discussed earlier. Interestingly, Mattila and Seppälä [79] understand a blockchain network in terms of a multisided platform and demonstrate how platform developers, application providers, users, miners, and other network participants are subject to different types of incentives and fulfill different roles as regards platform innovations (software update proposals) and other decisions affecting the evolution of the blockchain network. One could expect that for different blockchain platforms this will turn out differently, with different implications for their innovation capability.

Last but not the least, blockchain in itself is not just a digital technology or infrastructure or network which requires governance, but in itself it constitutes a model of governance as distinct from markets and hierarchies [80, 81]. As such, it underlies organizational forms that are autonomous and decentralized. Such organizational forms potentially are capable to innovate. For example, it seems that specific types of blockchain platforms such as initial coin offering (ICO) can become an indirect accelerator of innovation through creation of innovative blockchain start-ups. In an ICO, tokens of a blockchain start-up are offered to investors to fund the start-up based on cryptocurrencies. Even Ethereum originally was created in 2014 through an ICO, and currently is the leading blockchain platform for ICOs.

The blockchain ecosystem as described by Tapscott and Tapscott [74] could be considered as the environment in which blockchain innovations are initiated and implemented. This should not imply that there is no diversity of ecosystems. Actually, blockchain emerged as a commons based peer production activity [78]. An interesting overview of "blockchain innovation commons" is provided by Allen [82], who describes a variety of instruments such as fairs and conferences, embassies and centers, and forums and message boards. These are considered as private governance mechanisms facilitating the entrepreneurial process of discovering uses for blockchain technology.

13.7 Conclusions and Outlook

This chapter has presented digital innovation as a multidimensional concept. Digital innovation comprises both the object and the process of innovation, which affects both the innovation environment and mechanisms to steer and manage innovation processes. As digital innovations are systemic and interoperable in nature, the process of digital innovation is increasingly situated in platform ecosystems rather than within the company boundaries or restricted open innovation settings. Furthermore, digital platforms and tools inescapably become intertwined with the innovation process, as they provide the environment of innovation and support the interactions, collaborations, and engineering activities. This implies that the digital innovation process gets shape as a result of (a) the nature of the innovation as an object (modular, systemic, interoperable), (b) the innovation environment (increasingly, the platform innovation system), and (c) the governance and orchestration of digital innovation (based on rules and roles aligned to the innovation environment).

Implications of digital innovation for organizations and management received a lot of attention. Digital technologies and decreasing costs of communication enable decentralized forms of collaboration and organization. As a result, wide spectrum of smart organizations and collaborative networks has emerged and more and more attention has been paid to how collaborative workplaces and employment conditions may stimulate learning and innovation. Furthermore, and given the nature of digital innovation, these decentralized forms of organizing also help creating platform-based mechanisms to organize the innovation process. In addition, digital web-based tools for innovation have matured, such as digital marketplaces, idea generation, cocreation, crowdsourcing, knowledge sharing, collective intelligence, innovation prizes, and funding platforms.

These trends and forces work out differently according to the situational context (Table 13.2). Three areas of innovation have been

Table 13.2   Digital innovation in three societal contexts.

Digital innovation aspects	Companies and supply chains	Urban and regional development	Blockchain organizations
Focus of digital innovation	Products, services, processes, business models, workplaces, organizations, supply chains, enterprise networks	Public services, service platforms, public infrastructures (transport, energy, water, e-government), social innovation, citizen engagement	Secure, "trustless" peer-to-peer transactions; smart contracts; and decentralized autonomous organizations
Digital innovation environment	Industry competition; cooperation in supply chains; supply networks, platform ecosystems, learning factories, and crowdsourcing platforms	Urban social and economic challenges; political and business actors and goals; community participation, urban living labs, innovation competitions, and crowdsourcing platforms	Public and private blockchain platforms, consortium platforms, and indirectly also crowd-funding platforms such as the ICOs
Governance of digital innovation	Open innovation business models, partner selection, platform decision rules, platform leadership, shared values, and shared benefits	Public–private cooperation agreements; program-based innovation projects, and rules for innovation tendering and competitions	Blockchain protocols as governance, voting rules, blockchain community vision building, mechanisms to set up ICOs, and intellectual property rights management

addressed in this chapter to illustrate how digital innovation is shaped: industrial innovation, urban innovation, and innovation in blockchain organizations. Industrial digital innovation transforms from "traditional" open innovation limited to few companies to network- and ecosystem-oriented forms open innovation, driven by customer-driven supply chains, product-as-service business models, and industrial platforms. The growing attention for workplace innovation as a condition for success of digital transformation and concepts such as the "learning factory" may balance the technology-oriented thinking of Industry 4.0. Different types of industrial innovation environments such as small-scale social and entrepreneurial initiatives, regional ecosystems, and business supply chains will be coexisting and mutually enforcing.

Urban digital innovation seems to have crossed the technocratic phase of smart city thinking. As platforms and social networks shape "smart city" transformations, the challenge is now to integrate urban innovation and its wide-ranging economic, social, legal, and political implications within the democratic process. Cities and regions have become important environments of social and business innovation and entrepreneurial ecosystems based on smart specialization strategies and local challenges. Platform environments, living lab settings, and availability of easy-to-use digital innovation tools in combination with community engagement and social innovation are promising developments to shape and implement urban innovation environments. More radically decentralized organizing models such as peer production and blockchain strongly affect our thinking about decentralized ways of working, production, and innovation. The peer-to-peer model does not yet seem to suit industrial contexts, apart from software and content production and small-scale forms of open manufacturing. The prospects of blockchain networks to stimulate innovation probably will be limited to processes that can be automated through blockchain technologies, such as setting up the ICOs and managing intellectual property rights.

In the three domains studied, traditional phase-oriented forms of innovation are replaced or at least complemented with ecosystem models of innovation which require appropriate models of trust building, governance, and orchestration. Currently evolving high-impact developments such as the COVID-19 pandemic combined with climate change challenges are creating a lot of uncertainty but also radically new opportunities as regards decentralized forms of organizing, production, work, and innovation.

References

[1] OECD/Eurostat (2018). "Oslo Manual 2018: Guidelines for Collecting, Reporting and Using Data on Innovation", 4th Edition, "The Measurement of Scientific, Technological and Innovation Activities", OECD Publishing, Paris/Eurostat, Luxembourg. https://doi.org/10.1787/9789264304604-en. Date of access: April 14, 2020.

[2] Tidd, J. and Bessant, J. (2009). "Managing Innovation. Integrating Technological, Market and Organizational Change". 4th Edition, John Wiley & Sons Ltd, Chichester, England.

[3] Chesbrough, H., Gassmann, O. and Enkel, E. (2010). "The future of open innovation". R&D Management Special Issue: The Future of Open Innovation. 40(3), 213-221.

[4] Brunswicker, S. and Vanhaverbeke, W. (2015). "Open innovation in SMEs: External knowledge sourcing strategies and internal organizational facilitators". Journal of Small Business Management, 53, 4, 1241-1263. DOI: 10.1111/jsbm.12120

[5] Lundvall, B.-Å. (2010). "National Systems of Innovation: Toward a Theory of Innovation and Interactive Learning". Anthem Press, London. (First edition, 1992).

[6] Etzkowitz, H. and Leydesdorff, L. (2000). "The dynamics of innovation: From national systems and 'Mode 2' to a triple helix of university-industry-government relations". Research Policy, 29, 2, 109-123.

[7] Linden, G., Kraemer, K. and Dedrick, J. (2009). "Who captures value in a global innovation network? The case of Apple's iPod". Communications of the ACM - Being Human in the Digital Age. CACM Homepage archive, 52, 3, 140-144.

[8] Mazzucato, M. (2013). "The Entrepreneurial State. Debunking Private vs. Public Sector Myths". Anthem Press, London.

[9] Wieczorek, A.J. and Hekkert, M.P. (2012). "Systemic instruments for systemic innovation problems: A framework for policy makers and innovation scholars". Science and Public Policy, 39(1), 74-87.

[10] Porter, M.E. and Heppelmann, J. (2014). "How Smart, Connected Products are Transforming Competition". Harvard Business Review 92, 11, pp 64-88.

[11] Slama, D., Puhlmann, F., Morrish, J. and Bhatnagar, R.M. (2016). "Enterprise IoT. Strategies and Best Practices for Connected Products and Services". O'Reilly, Sebastopol.

[12] Mazzucato, M. (2018). "The Value of Everything. Making and Taking in the Global Economy". Penguin Books, UK.

[13] Zuboff, S. (2019). "The Age of Surveillance Capitalism". Profile Books, London.

[14] Miranda, J., Mäkitalo, N., Garcia-Alonso, J., Berrocal, J., Mikkonen, T., Canal, C. and Murillo J.M. (2015). "From the Internet of Things to the Internet of People". IEEE Internet Computing 19, 2, 40-47, March/April 2015, DOI: 10.1109/MIC.2015.24

[15] Gawer, A. and Cusumano, M. (2013). "Industry platforms and ecosystem innovation". Journal of Product Innovation Management, 31, 3, 417-433

[16] Tiwana, A. (2014). "Platform Ecosystems. Aligning Architecture, Governance and Strategy". Morgan Kaufman, Waltham.

[17] Malone, T. (2004). "The Future of Work". Harvard Business School Press, Boston, Massachusetts.

[18] Benkler, Y. (2007). "The Wealth of Networks. How Social Production Transforms Markets and Freedom". Yale University Press, New Haven and London.

[19] Boorsma, B. (2017). "A New Digital Deal. Beyond Smart Cities. How to Best Leverage Digitalization for the Benefit of Our Communities". Rainmaking Publications. www.anewdigitaldeal.com

[20] Yoo, Y., Henfridsson, O. and Lyytinen, K. (2010). "A new organising logic of digital innovation: an agenda for information systems research". Information Systems Research, 21, 4, 724-734.

[21] Yoo, Y., Boland, R.J., Lyytinen, K. and Majchrzak, A. (2012). "Organizing for innovation in a digitized world". Organization Science, 23, 5, September–October, 1398-1408. Special issue introduction.

[22] Nambisan, S., Lyytinen, K., Majchrzak, A. and Song, M. (2017). "Digital innovation management: reinventing management research in a digital world". MIS Quarterly, 41, 1, 223-238. Special issue introduction.

[23] Nambisan, S., Wright, M. and Feldman, M. (2019). "The digital transformation of innovation and entrepreneurship: Progress, challenges and key themes". Research Policy, 48. https://doi.org/10.1016/j.respol.2019.03.018.

[24] Autio, E. and Thomas, L.D.W. (2014). "Innovation ecosystems: Implications for innovation management". In "The Oxford Handbook of Innovation Management," Mark Dodgson, David M Gann, Nelson

Phillips (eds.), Edition 1, Chapter 11, Oxford University Press, Oxford, pp. 204-228.

[25] Thomas, L.D.W. and Autio, E. (2020). "Innovation ecosystems in management: An organizing typology". In "Oxford Encyclopaedia of Business and Management". Oxford University Press, Oxford. DOI: 10.1093/acrefore/9780190224851.013.203

[26] Macedo, P. and Camarinha-Matos, L. (2017). "Value systems alignment analysis in collaborative networked organizations management". Applied Sciences, 7, 1231. Retrieved: www.mdpi.com/2076-3417/7/12/1231/pdf. DOI: 10.3390/app7121231. Date of access: April 24, 2018.

[27] Isckia, T., De Reuver, M. and Lescop, D. (2018). "Digital innovation in platform-based ecosystems: An evolutionary framework". In Proceedings of the 10th International Conference on Management of Digital Ecosystems (MEDES'18) September 25–28, 2018, Tokio. ACM, New York, NY, USA.

[28] Nylén, D. and Holmström, J. (2015). "Digital innovation strategy: a framework for diagnosing and improving digital product and service innovation". Business Horizons, 58, 57-67.

[29] Svahn, F., Mathiassen, L., Lindgren, R. and Kane, G. (2017). "Mastering the digital innovation challenge". MIT Sloan Management Review, 58, 3, 14-16.

[30] Kane, G., Palmer, D., Phillips, A.N., Kiron, D. and Buckley, N. (2019). "Accelerating digital innovation inside and out". MIT Sloan Management Review and Deloitte Insights, June, 1-30. Date of access: January 8, 2020.

[31] Arundel, A., Lorenz, E., Lundvall, B.-A. and Valeyre, A. (2007). "How Europe's economies learn: A comparison of work organization and innovation mode for the EU-15". Industrial and Corporate Change, 16, 6, December, 1175-1210, https://doi.org/10.1093/icc/dtm035. Date of access: January 6, 2020.

[32] Oeij, P.R.A., Rus, D. and Pot, F. (2017). "The way forward with workplace innovation". In Oeij, P., Rus, D. and Pot, F. (eds.), "Workplace Innovation. Theory, Research and Practice". Springer International Publishing, Cham, 399-410.

[33] Kopp R., Howaldt J. and Schultze J. (2016). "Why Industry 4.0 needs workplace innovation: A critical look at the German debate on advanced manufacturing". European Journal on Workplace Innovation 2, 1, 7-24.

[34] Pot, F., Dhondt, S., Oeij, P., Rus, D. and Totterdill, P. (2019). "Complementing digitalisation with workplace innovation". In J.

Howaldt, C. Kaletka, A. Schröder and M. Zirngiebl (eds.), "Atlas of Social Innovation – 2nd Volume: A World of New Practices", (pp. 42-46). Oekoem Verlag, München.

[35] Adner, R. and Kapoor, R. (2010). "Value creation in innovation ecosystems: How the structure of technological interdependence affects firm performance in new technology generations". Strategic Management Journal, 31, 306-333.

[36] Musiolik, J., Markard, J. and Hekkert, M. (2012). "Networks and network resources in technological innovation systems: Towards a conceptual framework for system building". Technological Forecasting & Social Change, 79, 1032-1048.

[37] Provan, K.G. and Kenis, P. (2007). "Modes of network governance: Structure, management and effectiveness". Journal of Public Administration Research and Theory, 18, 2, 229-252.

[38] Camarinha-Matos, L.M. and Afsarmanesh, H. (2018). "Roots of collaboration: Nature-inspired solutions for collaborative networks". IEEE Access, 6, 30829-30843. DOI: 10.1109/ACCESS.2018.2845119. Date of access: July 22, 2018.

[39] Oh, D.-S., Phillips, F., Park, S. and Lee, E. (2016). "Innovation ecosystems: A critical examination". Technovation, 54, 1-6.

[40] Eisenman, T., Parker, G. and Van Alstyne M. (2006). "Strategies for two-sided markets". Harvard Business Review, 84, 10, 92-101.

[41] Pellika, J. and Ali-Vehmas, T. (2016). "Managing innovation ecosystems to create and capture value". Technology Innovation Management Review, 6, 10, 17-24.

[42] Valkokari, K., Seppänen, M., Mäntylä, M. and Jylhä-Ollila, S. (2017). "Orchestrating innovation ecosystems: A qualitative analysis of ecosystem positioning strategies". Technology Innovation Management Review, 7, 3, 12-24.

[43] Selander, L., Henfridsson, O. and Svahn, F. (2010). "Transforming ecosystem relationships in digital innovation". In ICIS 2010 Proceedings, 138. http://aisel.aisnet.org/icis2010_submissions/138. Date of access: November 18, 2019.

[44] Kohli, R. and Melville, R. (2019). "Digital innovation: A review and synthesis". Info Systems Journal, 29, 200-223.

[45] Fernández-Macías, E. (2018). "Automation, digitisation and platforms: Implications for work and employment". Eurofound, Publications Office of the European Union, Luxembourg.

[46] Bauwens, M., Kostakis, V. and Pazaitis, A. (2020). "Peer to Peer. The Commons Manifesto". University of Westminster Press, London.

[47] Cox, A., Rickard, C. and Tamkin, P. (2012). "Work Organisation and Innovation". Eurofound, Publications Office of the European Union, Luxembourg.

[48] Industrial Innovation in Transition project website: https://www.iit-project.eu/deliverables/. Access date: August 30, 2017.

[49] Thoben, K.-D., Wiesner, S. and Wuest, T. (2017). "Industry 4.0 and smart manufacturing. A review of research issues and application examples". International Journal of Automation Technology, 11(1), 4-16.

[50] Skylar, A., Kowalkowski, C., Tronvoll, B. and Sörhammer, D. (2019). "Organising for digital servitization: A service ecosystem perspective". Journal of Business Research 104, 450-460, https://doi.org/10.1016/j.jbusres.2019.02.012. Date of access: May 16, 2019.

[51] Lenkenhoff, K., Wilkens, U., Zheng, M., Süsse, T., Kuhlenkötter, B. and Ming, X. (2018). "Key challenges of digital business ecosystem development and how to cope with them". Procedia CIRP, 73, 167-172. https://doi.org/10.1016/j.procir.2018.04.082. Date of access: March 13, 2019.

[52] Abele, E., Metternich, E. and Tisch, M. (2019). "Learning Factories. Concepts, Guidelines, Best-Practice Examples". Springer International Publishing, Cham.

[53] Bettencourt, L. (2013). "The kind of problem a city is". Santa Fé Institute Working Paper 2013-03-008.

[54] UN-Habitat (2016). "World Cities Report 2016: Urbanization and Development - Emerging Futures. UN-Habitat.

[55] Cassandras, C.G. (2016). "Smart cities as cyber-physical social systems". Engineering, 2, 156-158.

[56] Hollands, R.G. (2015). "Critical interventions into the corporate smart city". Cambridge Journal of Regions, Economy and Society 8, 61-77. DOI:10.1093/cjres/rsu011.

[57] Koolhaas, R. (2014). "My Thoughts on the Smart City". Retrieved: http://ec.europa.eu/archives/commission_2010-2014/kroes/en/content/my-thoughts-smart-city-rem-koolhaas.html. Date of access: April 7, 2019.

[58] Sanchez, L., Muñoz, L., Galache, J.A., Sotres, P., Santana, J.R., Guttierrez, V., Ramdhany, R., Gluhak, A., Krco, S., Theodorides, E.

and Pfistere, D. (2014). "SmartSantander: IoT experimentation over a smart city testbed". Computer Networks, 61, 217-238. Retrieved: https://doi.org/10.1016/j.bjp.2013.12.020. Date of access: January 28, 2019.

[59] Struye, J., Braem, B., Latré, S. and Marquez-Barja, J. (2018). "The CityLab testbed - Large-scale multi-technology wireless experimentation in a city environment: Neural network-based inference prediction in a smart city". In IEEE INFOCOM 2018 IEEE Xplore. DOI: 10.1109/INFCOMW.2018.8407018.

[60] Castelnovo, W. (2016). "Citizens as sensors/information providers in the co-production of smart city services". In Agrifoglio, R., Caporarello, L., Magni, M. and Za, S. (eds.). "Re-shaping Organizations through Digital and Social Innovation." Proceedings of the 12th Annual Conference of ITAIS, LUISS University Press, 51-62.

[61] Alvear, O., Calafate, C.T., Cano, J.C. and Manzoni, P. (2018). "Crowdsensing in smart cities: Overview, platforms and environmental sensing issues". Sensors, 18, 460. DOI: 10.3390/s18020460.

[62] Serrano, M., Isaris, N., Schaffers, H., Domingue, J., Boniface, M. and Korakis, T. (eds.) (2017). "Building the Future Internet through FIRE. 2016 FIRE Book: A Research and Experiment based Approach". River Publishers, Denmark/The Netherlands.

[63] Lanza, J., Sánchez, L., Muñoz, L., Galache, J.A., Sotres, P., Santana, J.R. and Gutiérrez, V. (2015). "Large-scale mobile sensing enabled internet-of-things testbed for smart city services". International Journal of Distributed Sensor Networks, 2015, Article ID 785061, DOI: 10.1155/2015/785061.

[64] Sotres, P., Santana, J.R., Sánchez, L., Lanza, J. and Muñoz, L. (2017). "Practical lessons from the deployment and management of a smart city IoT Infrastructure: The SmartSantander Testbed Case". IEEE Access, 5, 14309-14322. DOI: 10.1109/ACCESS.2017.2723659.

[65] Demeester, P., Van Daele, P., Wauters, T. and Hrasnica, H. (2017). "FED4FIRE – The Largest Federation of Testbeds in Europe". In Serrano, M., Isaris, N., Schaffers, H., Domingue, J., Boniface, M., Korakis T. (eds.), "Building the Future Internet through FIRE. 2016 FIRE Book: A Research and Experiment based Approach." River Publishers, Denmark/The Netherlands, 87-109

[66] SynchroniCity (2019). Press Release. Retrieved: https://synchronicity-iot.eu/wp-content/uploads/2019/02/SynchroniCity_060219-Press-release-.pdf. Date of access: May 19, 2019.

[67] Schaffers, H. and Turkama, P. (2015). "Research and Innovation Programmes Shaping Ecosystems for Open Innovation, Some Lessons". Open Innovation 2.0 Yearbook 2015. Retrieved: https://ec.europa.eu/newsroom/dae/document.cfm?doc_id=9637. Date of access: February 2, 2015.

[68] Schaffers, H., Boniface, M. and Kirkpatrick, S. (2016). "The Role of Experimentation Facilities in Open Innovation Ecosystems for the Future Internet". In Open Innovation 2.0 Yearbook 2016. Retrieved: https://ec.europa.eu/newsroom/dae/document.cfm?doc_id=16072. Date of access: June 19, 2016.

[69] Schaffers, H. et al (2017). "European challenges for experimental facilities". In Serrano et al (eds.), "Building the Future Internet Through FIRE: A Research and Experiment-Based Approach". River Publishers, Denmark/The Netherlands, pp. 3-42.

[70] Leminen, S., Rajahonka, M. and Westerlund, M. (2017). "Towards third-generation living lab networks in cities". Technology Innovation Management Review, 7, 11, 21-35.

[71] De Waal, M. and Dignum, M. (2017). "The citizen in the smart city. How the smart city could transform citizenship". Information Technology, 59, 6, 263-273.

[72] Caragliu, A. and Del Bo, C.F. (2019). "Smart innovative cities: The impact of Smart City policies on urban innovation". Technological forecasting and social change, 142, 373-383.

[73] Future Cities Catapult (2017). "Smart Cities Strategies, A Global Review". Retrieved: http://futurecities.catapult.org.uk/wp-content/uploads/2017/11/GRSCS-Final-Report.pdf. Date of access: March 24, 2019.

[74] Tapscott, D. and Tapscott, A. (2017). "Realizing the Potential of Blockchain. A Multi-Stakeholder Approach to the Stewardship of Blockchain and Cryptocurrencies". White Paper, World Economic Forum. Retrieved: http://www3.weforum.org/docs/WEF_Realizing_Potential_Blockchain.pdf

[75] Morabito, V. (2017). "Business Innovation through Blockchain". Springer International Publishing, Cham.

[76] Panescu, A.-T. and Manta, V. (2018). "Smart contracts for research data management over the Ethereum blockchain network". Science & Technology Libraries, 37(3), 235-245.

[77] Narayan, R. and Tidström, A. (2019). "Blockchains for accelerating open innovation systems for sustainability transitions". In

Swan, M., Potts, J., Takagi, S., Witte, F. and Tasca, P. (eds.), "Blockchain Economics: Implications of Distributed Ledgers: Markets, Communications Networks, and Algorithmic Reality", pp. 85–101. World Scientific Publishing Europe Ltd, London. https://doi.org/10.1142/9781786346391_0005. Date of access: March 23, 2020.

[78] Pazaitis, A. (2020). "Breaking the chains of open innovation: Post-blockchain and the case of Sensorica". Information 11, 104. doi:10.3390/info11020104.

[79] Mattila, J. and Seppälä, T. (2018). "Distributed governance in multi-sided platforms". In A. Smedlund, A. Lindblom and L. Mitronen (eds.), "Collaborative Value Co-creation in the Platform Economy". Translational Systems Sciences, No. 11, 183-205, DOI: 10.1007/978-981-10-8956-5_10.

[80] Davidson, S., De Filippi, F. and Potts, J. (2016). "Economics of Blockchain". Retrieved: SSRN: https://ssrn.com/abstract=2744751 or http://dx.doi.org/10.2139/ssrn.2744751. Date of access: April 12, 2018.

[81] De Filippi, P. and Wright, A. (2018). "Blockchain and the Law: The Rule of Code". Harvard University Press, Cambridge, Massachusetts.

[82] Allen, D.W.E. (2017). "Blockchain innovation commons". SSRN Electronic Journal, available at https://papers.ssrn.com/sol3/papers.cfm?abstract_id=2919170 DOI: 10.2139/SSRN.2919170. Access date: May 8, 2020.

14

Open and Cooperative Infrastructures for Commons-Based Economies

Michel Bauwens[1] **and Sarah Manski**[2]

[1]P2P Foundation, Amsterdam, The Netherlands
[2]George Mason University, Fairfax, VA, USA

Abstract

This chapter discusses a new model of value creation called "commons-based peer production" (CBPP). The authors discuss whether it offers new solutions for integrating externalities in our economic systems. CBPPs are open, collaborative ecosystems that allow for a fluid flow of contributions toward the joint construction of common goods, i.e., the commons. The authors define the commons as shared resources that are maintained or produced by a community or a group of stakeholders, governed according to the rules and norms of that community. The chapter elaborates on the forms of cooperation enabled by commons-centric economies, and presents a strategy for commons-based value creation and capture. It also discusses in detail the relation between commons-based economies and blockchain networks.

Keywords: Commons-based peer production, value creation, distributed ledger, blockchain, holochain

14.1 Describing the Context

14.1.1 Commons-based Peer Production and the Need for Generative Market Forms

In this chapter, we will look at the emergence of a new model of value creation, which Yochai Benkler [1] has called "commons-based peer production" (CBPP), and whether it offers new solutions for integrating

319

externalities in our economic systems. CBPPs are open, collaborative ecosystems that allow for a fluid flow of contributions toward the joint construction of common goods, i.e., the commons. We define the commons as shared resources that are maintained or produced by a community or a group of stakeholders, governed according to the rules and norms of that community.

The first modern iteration of the CBPP model was the production of digital goods, or the joint production of knowledge, software, and designs that are commonly accessible via digital networks. Of course, the production of such goods requires material infrastructures and human bodies, but the output is considered nonrival or even antirival[1] [2] because it can be reproduced digitally. This means that the resulting production either does not lose value by being shared or that it actually gains value the more it is used. After the initial extraction of value from nature and human labor, the nonrival product's duplication requires very little additional input. Importantly, these digital resources have value among a community of users whether or not they have "market" or exchange value. These digital goods' value is in the first place their "use value," but since these resources are abundant, there is not direct creation of market value, which requires scarcity; a tension between supply and demand. However, they can be a commons for a commercial sector, and thus generate markets that use them, surround them, and create added value through them. What is innovative in peer production, is that the primary motivation is the creation of use value.

A report from 2011, Fair Use in the U.S. Economy [3], calculated the value for the US economy of activities centered around such shared resources to be one-sixth of GDP. These figures reflect the situation before the growth of another form of—often by private platforms—mutualization in the form of shared services, the so-called "sharing economy." The concept of sharing does not here denote sharing as classically understood, but rather idle-sourcing, defined as the capacity to put resources such as housing and transportation into a common usage pool. These resources may have previously been dormant or were difficult to bring to market before the ease of use of digital networks. In this latter form, while the platforms are generally intended to create market exchange between peers directly, there

---

[1]Steven Weber, the author of the landmark book, The Success of Open Source, argues that it is not enough to characterize free software as a nonrival resource. It is more than antirival, because it benefits from network effects: more use, even "use for free," is beneficial.

are no commons or really shared resources. However, there is an emergent part of the sharing economy that practices "platform cooperativism" [4]. In this model, it is the platform itself that is the shared resource, as it maybe commonly held by an association, a cooperative, a group of users, or collectively managed by a group of stakeholders. In this context, it is the ecosystem that functions as the common infrastructure for a particular market that functions as a commons, because it is not privately owned or managed.

14.1.2 Common Structures Found in Commons-based Peer Production

After 10 years of research by the P2P Foundation and its associated P2P Lab, we believe that most CBPP efforts have a common structure, which we have since confirmed with a study of urban commons in the city of Ghent in 2017 [5, 6] which focused on citizen initiatives around shared resources and common provisioning systems. There are three shared institutions involved in the CBPP ecosystems.

The *first* core institution of the CBPP ecosystems is *the open productive community*; a platform in which contributors can permissionlessly offer their contributions; meaning anyone can join the network. It is through this self-organized community that individuals can add contributions to the common good. This defines their project through a common set of rules and norms, governing the contribution, selection, and "maintainer" processes for these open source communities. The value regime of CBPP is not commodity labor, but all contributions, whether they are commodity wage labor, paid freelance work, or freely offered contributions. In an open productive community, nonwaged labor can indeed freely participate in the coconstruction of the commons. Overall, the distribution of tasks can be coordinated outside of the hierarchical control of participating firms, because open systems are holoptical[2]. This means that the full production system can be seen, and coordination occurs through stigmergy

---

[2]Peer groups are characterized by holoptism, i.e., the ability for any member to have horizontal knowledge of what the others are doing, but also the vertical knowledge related to the aims of the project: "Holoptism is the implied capacity and design of peer to processes that allows participants free access to all the information about the other participants; not in terms of privacy, but in terms of their existence and contributions (i.e., horizontal information) and access to the aims, metrics, and documentation of the project as a whole (i.e., the vertical dimension)" [7].

[8]: free mutual coordination based on the availability of signaling between participants.

The *second* institution involved in the CBPP ecosystems is the marketplace that uses and surrounds the commons to realize or add market value through their participation in these commons. This "entrepreneurial coalition," [9] consisting of firms and marketable labor, adds commodity value to the shared commons by constructing an ecosystem of added products and services that can be sold in the market. Those commoners and contributors that want and need to make a living through their contributions will need to join the market (or employment in the state sector) to create livelihoods related to their contributory activity, while more classic actors such as firms will seek to create profit out of their relation to these commons. What matters here is the quality of the relationship between the commoners and their open productive communities, and the market players and ecosystem. The key question to be answered for every project where this cooperation between commons and markets occur is the following: Is the relationship an extractive one, whereby market players exploit and weaken the commons and their contributors, or is it a generative relationship, in which the market activities strengthen and expand the commons, while creating good livelihoods for the commoners? There are various attempts to create these new generative forms, such as the "shared enterprise model" pioneered by Smart.coop, or the "multifactory model" [10]. Majorie Kelly, in her book on the *Emerging Ownership Evolution* [11] has developed five criteria for the generativity of ownership and governance models that work for the commons.

The *third* institution involved in the CBPP ecosystems is what we call the "for-benefit association," which often takes the form of the so-called FLOSS Foundations in the open source economy [12]. An open productive ecosystem needs a common infrastructure that is not controlled and owned entirely by individual market participants, and this infrastructure needs to be maintained and expanded. The majority of CBPP projects therefore have their infrastructure of cooperation governed by a separate institution, "which enables and empowers cooperation." Unlike classic NGOs, these do not operate from the basic idea of scarce resources that need to be directed through a market pricing or command and control mechanism, but instead maintain open platforms over time, so that they can be used by all contributors. The core principle is the collaborative platform will attract contributions from a theoretically abundant pool of potential collaborators interested in those particular projects. These foundations also manage the relationship between the productive community and the private players, maintain a common culture and knowledge base (education, meetings, etc.), defend shared and open knowledge (intellectual property for the commons), and create joint certification, training, and other joint activities. An example would be the role of the Wikimedia Foundation vis-à-vis Wikipedia, or the Linux Foundation, Drupal Association, and other FLOSS institutions.

CBPP ecosystems are a modern manifestation of the desire to organize production around the commons. At the microeconomic level, the relationship between the commons and the market has a long history. One good source for that history is the recent essay of Adam Arvidsson, Capitalism and the Commons [13], and the book, *Changemakers* [14], where Fay Weller and Mary Wilson argue the commercial revolution in early medieval cities in Europe in the period starting in the 12th century was largely determined by a strong link between commons-forms and the bottom-up markets it enabled in the re-emerging cities. Those that left the countryside created urban commons and guild systems. These functioned as social solidarity systems for urban workers, and created "moral markets" that would be theorized by the Italian scholars of the 13th–14th century, while in the regions marked by rural freeholders, there was a surge of contractual rural commons (see also [15]). So, already in that period, we see a tight coupling between commons forms and market forms.

In this post-2008 historical juncture, which can also be interpreted as a time of transition, Arvidsson argues the crisis of capitalism is creating a similar dynamic: both the declining middle classes of the West, and the migratory informal workers in the Global South, are creating "capital-poor but labor-intensive" economic forms, which he calls "industrious capitalism." This type of capitalism very strongly relies on different kinds of commons without which it would not extract profit. Alain Tarrius describes this as a "peer-to-peer, poor-to-poor" nomadic economy of rotative, transmigrant communities [16]. Arvidsson [14] discusses the Bangkokization of the world including the economies of the precarious and "cognitive" nomadic working class, and the network of places where they congregate such as Chiang Mai, Thailand, Ubud, Bali, Medellin, Colombia, and Tenerife.

14.2 Phenomenology of the Commons Economy

Figure 14.1 suggests that the emerging commons-centric economy is taking four important forms, which are all growing and struggling for their place in a

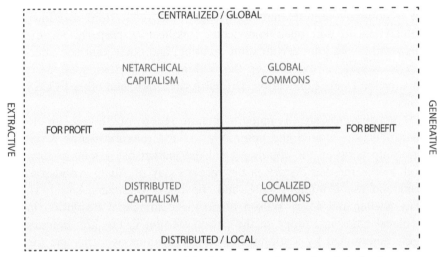

Figure 14.1   The four-quadrant model of four competing sociotechnical value systems. Graphic by Michel Bauwens [17], used with permission[3].

new ecosystem. The *first* form, *netarchical capitalism*, combines peer-to-peer access at the front end, but centralizes platform ownership, profit, control and user data, the peer-to-peer exchange of services (Uber, Airbnb), or the peer-to-peer sharing of knowledge (Google, Facebook), extracts profit by directly siphoning rent from the activities of the users. Value here is minimally related to paid commodity labor and maximally related to the independent exchange of the user base. Note that the Google/Facebook models have almost no rewards of benefit-profit sharing for their user base, while Airbnb/Uber create their profit out of the direct activities of their user base as well. In other words, they directly exploit human cooperation. This is "Proudhonian capitalism" since their profits are only marginally related to their use of commodified labor, but rather come from "free labor" or peer-to-peer exchanges that are "taxed."

The *second* form, which we call *distributed capitalism*, is also market oriented but aspires to "really free" markets outside the control of monopolies, and big institutions, including the state. The prime example of this is Bitcoin as a currency and the multitude of emerging projects surrounding this distributed blockchain ledger [20]. The concept and practice of a shared and distributed ledger has now moved beyond

---

[3]This framework was first introduced in Ref. [18] and was reworked and published in the form used in this chapter in Ref. [19].

the first version of this technology (blockchains) to other distributed ledger technologies, such as the Holochain project. In this distributed-capitalist model, where the software is open source, the community dynamics arc important, and the production model allows for permissionless contributions. With the bitcoin blockchain, the libertarian ethos of its original designers, following "Austrian Economics"[4] [21], is embedded through many design decisions regarding its concrete infrastructure, and the aim of the activities are market exchange and profit-making. Hence, while the blockchain ecosystems are open and collaborative, they are based on extractive, environmentally degrading technology. However, distributed ledger technology comes in many forms in this early "pluralistic environment," and just like private platforms have cooperative alternatives, so too do we see similar alternatives arising in the blockchain and post-blockchain space.

The other two quadrants, the *global commons and localized commons*, are represented by projects that do not have an explicit for-profit orientation, but instead a for-benefit one, i.e., a social purpose. In this model, eventual surplus and profit can be mobilized for the mission or purpose of the collective project. We distinguish the local, mostly urban commons projects, which mutualize various provisioning systems, from global infrastructures, which are deterritorialized. In the global model, we count the global open design, knowledge, and free software communities.

Each of these forms can be used, or transformed, to become a vector of an expanding commons-centric economy that acts to conserve natural resources. These both propose a set of transformative strategies, which we shall now explore in depth (Fig. 14.2).

14.3 Cooperative Forms for a Commons-centric Economy

14.3.1 Transforming Rent-seeking Private Platforms into Platform Cooperatives

Netarchical capital designs spaces for commoning and peer-to-peer exchanges that are outside of the control of the user base, and from which rent or transaction fees can be extracted. The alternative here is to

---

[4]See the discussions on the P2P Foundation wiki: (1) https://wiki.p2pfoundation.net/Bitco in_and_the_Blockchain_Are_Firmly_Anchored_in_Anarcho-Capitalist_Visions_of_a_Hyp er-Capitalist_Society and (2) https://wiki.p2pfoundation.net/Most_Cryptoeconomics_Do_N ot_Challenge_Neoclassical_Premises

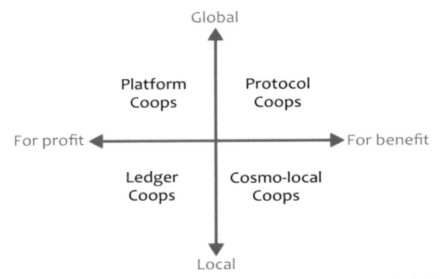

Figure 14.2   Summary of cooperative formats for a commons-centric economy (used with permission, graphic by Michel Bauwens [17]).

design the platform as a commons, i.e., to take control of the ecosystem or infrastructure of cooperation. The platform cooperative movement has already organized several global conferences and counts amongst its members, labor union financed platforms for domestic workers and nurses. The platform infrastructure can be owned by its workers, users, or a multistakeholder combination. One example is smart.coop. As a platform, it sits between freelancers, or autonomous workers, and the demand for their work by businesses or institutions. It takes care of the factoring (invoicing), and a 2% fee funds a mutual guarantee fund, which allows freelancers to be paid within a week after approval of their invoices by their clients. Another 4.5%, the factoring fee, is used to fund the staff and an ecosystem of services that continues to improve over time. After a set time, invoices are averaged into a legal salary, for the member who becomes formally an employee of the coop, which he co-owns, with taxes paid and guaranteeing the kind of protection that is normally reserved for salaried workers. In this model, a worker keeps his/her freedom to choose projects and clients, yet the risk has been socialized to the cooperative, which, having no shareholders, can continually reinvest in the common services. They have 20,000 cooperative members in 9 European countries.

14.3.2 Broadening the Scope of Urban Commons in the Context of Protocol Cooperatives

Our study in elaborating the commons transition plan for the city of Ghent [6] showed a tenfold increase in the number of urban commons projects in the city since 2008, and other studies by Oikos [22] have shown similar rates of growth for Flanders, Belgium, the Netherlands, and Catalonia. Iaione [23] has a database with 1,000 of those projects, half of them in the Global South, while other studies have shown similar dynamics in East Asia [24]. While these urban commons are effectively transforming urban provisioning processes as commons-based ecosystems, they are also underfunded, and rely mostly on precarious and voluntary labor. Hence the contradiction is that though these projects reduce ecological footprints and create community, they are also still marginalized. Manzini [25] described them as both small and local, while also open and connected. This cosmo-local aspect can be greatly strengthened through the use of public-commons protocols and policies. For example, the Bologna Regulation for the Care and Regeneration of the Urban Commons, copied and adapted in 200+ other Italian cities, has mobilized 800,000 Italians toward urban commoning. We specifically argue for the creation of "protocol cooperatives" in this specific context. These would be global, open design depositories, by a provisioning system (e.g., shared habitat, shared mobility), that would be underwritten by coalitions of cities, ethical finance, and representatives of generative economic sectors. These, we call "protocol cooperatives," as the common protocols are shared and scaled globally, creating "economies of scope"[5]. Thus, local projects would not have to constantly reinvent the wheel with their low capital base, but could rely on the adaptation of pre-existing free software packages that help the development of mutualized models of provisioning. One step further is that urban commons, which are now mostly "redistributing" resources that have been produced by the capital system, could themselves become productive actors, through the adaptation of cosmo-local models of production. In this model, "what is light is global and shared" and "what is heavy is local wherever possible." It combines relocalized distributed manufacturing, with globally shared knowledge bases. Ideally, we would like to see

---

[5]Economies of scope are best explained with the simple adage: 'doing more with the same input', each local point of production has access to globally shared and updated knowledge, which can be used at the point of production. For a discussion, see https://wiki.p2pfoundation.net/Economies_of_Scope

combinations of (1) global open depositories, (2) relocalized manufacturing, and (3) generative business models. These mutualized infrastructures have the capacity to dramatically reduce ecological footprints, are cheaper and more inclusive as they use fewer resources, while they maintain complex social systems, and are a school for self-governance in the productive sphere as well.

14.3.3 Strengthening the Transnational Scope of Open Design Communities through Open Cooperativism

Currently, open design communities, such as free software, are mostly inserted into neoliberal or capital-based ecosystems, such as Linux and are linked to big companies such as IBM. The relations are not necessarily generative for the free software developers. Blockchain ecosystems are one attempt to better distribute the surplus value created by these workers, as the token-based funding systems are distributed, and often 40% of the tokens are reserved for the work. But the blockchain system does not sufficiently transform the economy toward a generative commons. And the old-style cooperative and solidarity economy also have limitations. Needing to survive in a dominant capitalist system, and having to use double entry bookkeeping with limited access to capital, they nearly always evolve (the law of transformation), toward managerialism and "worker capitalism." The pressures of capitalism force an extractive enterprise on the outside, even if they are more democratic internally. Thus, the devolution of cooperatives into pragmatic market enterprises, that use the same competitive logic as the private sector, has been largely the general rule for the evolution of cooperatives. We should also note that in some cases private models evolve toward generative models, such as social entrepreneurship, as in the case of B-Corporations.

Our proposal to create a stronger linkage between the broader commons and the cooperative model is for the creation of "open cooperatives." An open cooperative is a productive entity that is formally organized to be generative rather than extractive, thus statutorily obliged to contribute common goods to the wider community and humanity. The aim of such a statute is to ensure that even as the law of transformation occurs, the commitment to the commons remains a legal obligation. This can be accomplished by contributing to the public or common good in some concrete way.

14.4 The Suggested Solution Space: Techniques for Reverse Cooperation

Here we want to present a stylized strategy for commons-based value creation and capture. We answer the question: how can a commons-community, and its surrounding ecosystem of generative enterprise, create alternative and more sustainable economic cycles outside of the dominant extractive system?

So, the first step is the creation of an open contributive community. This community declares "value sovereignty," which is a commitment to honor all membership contributions, and not just commodified work recognized by the external market. This can be concretized by the use of a contributive accounting mechanism, such as the one employed by Sensorica [27]. We also recommend the use of social charters, as in-group constitution, that derive its cooperative practice from its principles. The second step is the creation of a membrane between the internal predistribution, and the external sources of income, which are redistributed according to the new criteria. This requires a commitment to let at least a part of the external flow of currency payments, into a common fund that can be redistributed according to the contributive tokens or accounting system.

If capitalism is mostly about the privatization and extraction of value through surveillance and manipulation [28] from what was previously common, then a commons-centric economy does the opposite as it uses and transforms capital into the commons. This can be called transvestment, or the transformation of one form of value into another such as capital into commons, i.e., the opposite of investment, which is using capital to obtain more capital. Various techniques can be used to garner this effect, such as the capped returns approach, which was used by Enspiral to fund the development of Loomio.

We also recommend the use of new types of licenses that regulate the use of the commons such as copyfair. Whereas copyright privatizes and commodifies the use of knowledge, making sharing impossible, and copyleft allows even noncontributing extractive companies to use the shared knowledge base without reciprocity, copyfair splits knowledge sharing from commercialization. In this model, knowledge can be shared with everyone, but if commercialization is desired, reciprocity to the common effort is required. Finally, a nonprofit foundation or association, will be set up to manage the infrastructure of cooperation, which may own some machinery, defend against the abuse of the commons and the licensing scheme, organize certification, and also mediate conflicts, seeking common ground between the various externalities.

14.4.1 Accounting for Externalities

We believe that the value logic of capitalism does not sufficiently (if at all) integrate external thermodynamic constraints, nor negative social outcomes, while at the same time refusing to recognize the market value and positive outcomes generated by nonmarket activities. Market prices are realized through the tension between supply and demand. Commodity value is created by manufacturing and selling commodities, and this value is subject to taxation which can be used for redistribution. But there is no direct recognition of the cost of externalities, whether negative or positive, social or environmental. This means that a huge amount of generative activity cannot structurally be financed.

For example, the negative social externalities created by platform models such as those of Uber and Airbnb, are not taken into account in the corporate ledgers. For example, Airbnb has a huge impact on gentrification trends. The positive social externalities created by human sharing and exchange on communication platforms such as Facebook, without which it could not exist, are not rewarded and funded; Facebook would not have any value without the unpaid communication activities of its 2 billion users. The negative ecological externalities such as fertilizer pollution are not accounted for in traditional industrialized agriculture and the positive ecological externalities, such as reduction of pesticides through the activity of organic farmers, remain unrecognized and unfunded.

14.4.2 The Historical Importance of Accounting

Accounting is not a mere technical tool, but a tool that regulates the transformation of the external world. It accompanies major societal transitions and is linked to the dominant value logic of the socioeconomic system. The first writings in Sumer/Mesopotamia were ledgers indicating the comings and goings of goods in the temples, which was also a signpost to the emergence of state-like institutions standing outside of the direct peer-to-peer relations between people in earlier social systems. The synthesis and invention of the double-entry bookkeeping by Fra Luca Pacioli co-evolved with the emerging market and protocapitalist system in the Italian cities of the Middle Ages. This kind of accounting values what comes in and out of the corporate or trading entity, and whether or not profit is retained excludes an ecosystemic view of the economy.

We see that today, with three new forms of accounting that are emerging in the developmental space of distributed ledgers, each share emerging post-capitalist characteristics, which include:

1. introducing contributions and not just commodity value,
2. monitoring flow while abolishing double entry, and
3. introducing multicapital accounts for direct thermodynamic flow monitoring.

We consider this trend as coevolving with the emergence of a proto-postcapitalist system that has integrated ecological constraints in its logic of operation. Contributive accounting goes a long way toward solving the "value" problem of classic capitalism. Indeed, many activities that produce commons are at present voluntary and unpaid, while commoning also rests on reproductive and care work that remains unfinanced in the current system. Once we move to a contributive system, with value sovereignty, i.e., the productive community chooses autonomously what it values internally, we can also move to a pluralistic view of what value is, and encode it in predistributive systems. Predistribution means that we do not wait for the state to rectify social inequality "after the fact" through taxes (though that is still necessary) but that we immediately start distributing proceeds fairly within the commons-based productive ecosystems. This is already prefigured by tokenizing value in blockchain projects.

> "In my opinion, the currencies we know today are 'cold', and embody extractive, debt-based value systems. When we talk about Commons initiatives existing within these cold currency systems, as islands in a sea of capitalism, there are huge mismatches in value systems and exploitation of the commons is the norm. However, using a Cyber-Physical Commons framework, we can design 'warm' currencies that embody community value systems. We can imbue community currencies with semi-permeable membranes and governance toolkits (along with cultural best practices) that allow us to embed Pigouvian taxes to internalize externalities, and limit exploitation by capitalist systems".[6]

The Commons Stack value accounting mission is good people should be rewarded for doing good things. We can decry "profit incentives" as

---

[6]Jeff Emmett of the Commons Stack (e-mail to authors).

extractive in our "cold" monetary system, but in a "warm" monetary system it could alternatively be understood as a "generated value surplus." By designing commons ecosystems with positive feedback loops offering reciprocal rewards to people who provide value to the community, we are aligning the incentives of participants to propagate what is best for that community. This is preferable over expecting people to sacrifice their bank accounts to chase dreams of doing good in the world, only to quit and go back to the "real world" after realizing how little their efforts are financially valued.

In commons circles, it seems we are jaded by the existing financial system, and for good reason, but there is no reason to throw out the baby with the bath water. "Cold" currencies have a big issue with exploitation and value extraction, but with appropriate system design we can retain the useful aspect of incentive alignment in "warm" currency systems, and mitigate the extractive tendencies of unchecked capitalism. "Warm" currencies will allow us to step away from talking about "dollars" and start talking about "value." With community attribution networks to appropriately value care work and monitoring systems that respect complexity, we can address the shortcomings of our current system and build something new atop the rubble of neoliberalism.

Flow accounting addresses the systemic blindness of collective-ego oriented double entry bookkeeping. The resources-events-agents model means that in every transaction, two agents exchange (the event) a resource, and it models the totality of metabolic pathways in an ecosystem. Any member of the productive community sees the place of their actions directly in the ecosystemic flow of the production.

14.5 The Evolution of a Generative Blockchain Space

What kind of socio-technical infrastructure would we need to implement the global possibility for a socially just, and ecologically stable cosmo-local production infrastructure? To move to the next political economic system, we need to create technologies enabling social institutions that pull land, labor, currency, communication and governance out of global capitalism. Distributed ledger technologies have a large role to play in this process and the pace of the revolutionaries working toward a global technological commonwealth has been quietly progressing [29].

As is true of all innovations, distributed ledger technologies (DLTs) are built upon earlier technological innovations and collaborative problem solving. DLTs are the latest innovation in a long project by a community of digital privacy activists, or cypherpunks, to create technologies protecting privacy and a free society[7]. While we can interact with technologies, it is important to keep in mind that technological objects and systems have agency, which is closely aligned to the objectives of its creators. For those involved with DLT projects, the technological affordances allowing for a fundamental shift in value accounting, is inspiring a wave of activism designed to change the way our social, political, and economic societies work away from global capitalism to the commons. As actors engage with DLTs in this process of social and material coconstruction of the technology demonstrates its own agency as well, because a technology's form "calls forth" or enables or constrains different human actions, called material agency [30, 31, 32, 33].

Blockchain's affordances [34] could facilitate cooperative governance such as that found in Commons-Based Peer Production (CBPP) communities [35]. CBPPs are an emergent model of socioeconomic production in which groups of individuals cooperate with each other to produce shared resources without a traditional hierarchical organization [35]. Blockchain's seven affordances [34] include verifiability, globality, liquidity, permanence, ethereality, decentralization, and future focus (Table 14.1).

We are in a critical moment in which DLTs material agency can be turned toward the mutual benefit of the world's people or it can be systematically foreclosed by elite powerful global actors. It is vitally important that the community of developers, designers, investors, philosophers, and activists develop a democratic and socialist approach to the creation of a "cyber-physical commons" framework. Minor "green," "progressive," or "liberal" adjustments to the status quo will fall short of averting climate catastrophe. Without a coherent sociotechnical imaginary of the future we want to build, technologies' material agency reinforces existing power structures because those who currently maintain unequal power and resources are able to adapt the technology to their own purposes [36, 37, 38]. In this sense DLTs poses both utopian and dystopian possible futures. In the struggle to shape DLTs, major corporations will marshal vast lobbying resources to get restrictive laws passed and so we need to combine all our available resources with a powerful sociotechnical imaginary of a regenerative future. Whether, individuals, nation-states, corporations, technologists, or communities are empowered will depend heavily on the design choices that are made in the next few years and on the path dependencies, and political dimensions of

---

[7]See A Cypherpunk's Manifesto https://www.activism.net/cypherpunk/manifesto.html

Table 14.1   Seven affordances of blockchain technology and the structural qualities that produce them.

1. **Verifiability.** Transactions are assured through encrypted network consensus mechanisms in such a form that all transactions from the very first to the most recent are recorded in a ledger open to its maintainers, reducing information asymmetries.

2. **Globality.** Digital transactions and cultural information flows transcend geographic space and national borders.

3. **Liquidity.** Value liquidity is enhanced as the location of a store of value that does not depend on or is not under the direct control of a sovereign, central bank or private corporation.

4. **Permanence.** The ledger of a transaction is immutable by design.

5. **Ethereality.** Transactions are conducted in a digital medium.

6. **Decentralization.** The ledger is widely distributed among many stakeholders and maintainers.

7. **Future focus.** Found in newer developments of blockchain such as Ethereum, a stored autonomous self-reinforcing agency (SASRA) is formed in the temporal displacement of action through the use of smart contracts enabling the prefigurative recording of future transactions. Below the table write, "Reprinted with permission from Manski, Sarah G., and Manski, Ben R. (2018). No Gods, No Masters, No Coders? The Future of Sovereignty in a Blockchain World. Law & Critique, (2)29, 151–162.

the policies, practices, applications, and institutions created surrounding this technology.

As the material objects that make up our everyday life become the Internet of Things (IoTs), distributed ledger technologies will enable a pervasive lowering of transaction costs into even more areas of the economy. The IoTs will use distributed ledger technologies to process smart contracts and micropayments. For example, an internet connected charging station will automatically debit your electric self-driving car with a higher degree of reliability than the current system has been able to manage. When transaction costs drop past invisible thresholds, there will be sudden, dramatic, hard-to-predict aggregations and disaggregations of existing business models. We need to be ready with commons alternatives.

14.6 Commonizing the Blockchain Space

"We believe the answer to the underfunding of social goods and underserved community contribution is to reframe social goods as self-governing and continuously funded commons" [39].

The commons can be managed in a sustainable way by local communities of peers when communities communicate to build common protocols and rules that ensure their sustainability [40]. Distributed ledger technologies can be designed for the creation of self-sustaining commons economies where all participants profit according to the value that they produce rather than trying to conform to the capitalist economy. These are the "cyber-physical commons" powered by blockchain networks which are designed to align user incentives toward maintaining the network. Miners earn tokens, developers hold the tokens hoping their efforts will raise its value, and users purchase tokens creating demand and pay transaction fees.

For example, members of the Giveth team are using blockchain technology for good by building a toolkit for creating these new community economies. The project is called the Commons Stack, and is a collaboration with BlockScience, a Complex Systems Engineering R&D firm. The Commons Stack is using the emerging discipline of token engineering to design technological improvements to streamline community fundraising and decision-making, lowering the barriers for groups with shared goals to operate as distributed protocol cooperatives. They are doing this by producing design patterns for community toolkits, a library of code specifications, and reference implementations. These designs will be chain-agnostic and can be applied on data-centric and/or agent-centric architectures, although most developer interest so far exists in the Ethereum ecosystem, so that is likely where they will see their designs first implemented.

It is important to distinguish the concept of distributed ledgers from many current implementations of blockchain technology, several of which have structural and environmental issues that may or may not be overcome in future iterations. Hence, the global commons movement is paying close attention to post-blockchain ledgers, which have different underlying philosophies. For example, the Holochain distributed ledger does not aim for a single worldwide chain of transactions, in which every transaction needs to be verified with the total accumulating database of all global transactions. Instead, Holochain has a biomimetic philosophy, which allows for local and contextual open ledgers to connect with each other and become interoperable [41].

We believe that most blockchain-based ledgers represent anarcho-capitalist market values, i.e., neoliberalism on steroids, and that post-blockchain ledgers could reinforce, with appropriate and alternative design decisions, the commons-centricity and linkages to contributive and ecological value. Table 14.2 outlines some of the alternative principles for such commons-oriented ledgers. So, just as we can change platform capitalist platforms into polygoverned platform cooperatives; we can also adapt DLT

Table 14.2 Contrasting the propertarian blockchain with commons-based ledger systems (used with permission. Graphic by Michel Bauwens. Sourced from Ref. [17]).

LIBERTARIAN	vs.	COMMONS-BASED
Examples:		*Examples:*
Bitcoin, Ethereum, Blockchain		Holochain, Faircoin, EcSA
Principles:		*Principles:*
Commodity-Based Tokens and Cryptocurrencies		Mutual Credit, Contribution-Based, and Asset-Backed Tokens
Competitive Games		Cooperative Games
Smart Contracts *(individual to individual)*		Ostrom Contracts *(social contracts and charters)*
Oligarchic Proofs of Consensus *(one dollar, one vote)*		Distributed and Contributory Proofs
One World Ledger to Rule Them All *Blockchain*		Interoperable P2P Ledger Systems *Holochain*
Market Value		Value Sovereignty
Extractive Ecosystems		Generative, Nature-Friendly Ecosystems
Profit-Driven		Impact, Purpose, For-Benefit Driven
Trustless		Trustful (Web of Trust)

projects to make them more commons-centric or at least commons-friendly. These possible changes in the design of DLT systems are what Table 14.2 is intended to illustrate.

In Table 14.2, we are not attempting to flatten the discussion to a binary right/wrong of ledger architectures. While many applications to date in the blockchain space have been extremely disappointing, this does not preclude useful technologies being developed on data-centric architectures that may not be possible or appropriate for agent-centric systems. Both models are likely to be useful to serve specific ecosystem needs, with proper system modeling and design required in all cases. Token engineering is the key to well-designed ecosystems, rather than the particular architecture used.

Open shared ledgers are a key mutual coordination mechanism to shift open source coordination from software to manufacturing. Blockchain and distributed ledgers generally enable open and contributive ecosystem accounting (such as practiced by the Canadian Sensorica project [27], REA (resource - event - agent), which let us see flows in shared circular economies involving multiple players, and biocapacity accounting, which is based on direct vision of the flows of matter and energy. These types of contributory accounting systems promote fairness, openness, transparency,

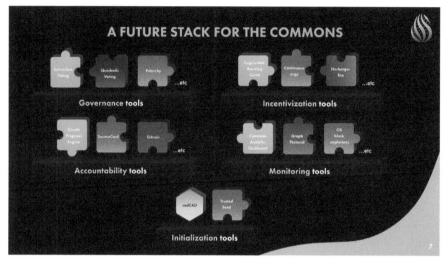

Figure 14.3 A future stack for the commons. The commons stack is building a library of tools for context-specific methods of governance, incentivization, accountability, monitoring, and initialization using holistic system simulations (Used with permission. Graphic by Jeff Emmet, published in "Commons Stack System Overview", forthcoming 2019).

security, and environmental limits. The current state of the blockchain world is one of fragmentation, but the tools are in development for the creation of interoperable P2P ledgers.

The Commons Stack is a project started in 2019 that aims to create community tools to improve decentralized coordination around shared goals. In these "community commons," blockchain technology is used to align economic incentives with each communities' values and scale these previously underfunded communal efforts into effective networks for good. They believe the growth of the commons will be accelerated through access to an open-source library of modular, customizable, and interoperable components enabling purpose-driven communities to unite around shared goals (Fig. 14.3).

The Commons Stack project has identified components for what they term a "minimum viable commons," to provide key functionality in coordinating a group around raising and allocating funds, making decisions, and measuring impact. The first component is the "augmented bonding curve" providing continuous funding for a commons initiative through community transvestment, with growing academic foundations for this new economic tool. The second is a transparent and accountable proposal service, which they call the "giveth proposal engine." The third is a novel process for continuous decision-making modeled off the mechanics of a neuron firing in the brain, called "conviction voting." And, finally, a means to monitor and measure the value produced in these communities, they term the "commons analytics dashboard," which they see leading to a future of computer-aided governance. The most important aspect of the Commons Stack is their emphasis on token engineering, including the use of an open source complex system modeling and simulation tool called cadCAD.

The Commons Stack could be the technological evolution needed to enable the growth of the commons by enabling crypto-economic systems of cooperation and governance. This modular "cultural and technical stack for the commons," could help communities reach shared goals by giving them the tools to bootstrap necessary funding (often a main hindrance to launch), and empowering that community with proportionally weighted peer governance, real time preference signaling, and monitoring systems that respect complexity. By creating a growing library of open source component blueprints for governance, funding, and other critical infrastructure, the Commons Stack enables communities to act as effective platform cooperatives, co-owning and comanaging shared funds as a commons. These components can be combined to create intentional, circular, and community-driven economies powered by continuous funding streams and transparent decision-making which will enable the threefold coordination of the post-capitalist economy (see the following sections).

14.7 Threefold Coordination of the Post-capitalist Economy

The 1930s saw a very vibrant "calculation debate" [42] between the disciples of Hayek and Schumpeter on the one side and the socialist "planning-oriented" left on the other. The socialist side of the debate should be understood in the context of the very rapid growth of the economy in the Soviet Union under Stalin, under the harsh conditions of coercive central planning, facing the arguments of the Hayekians, that distributed markets and their free price settings, were the only way to achieve a realistic coordination of a complex economy. Participants in this debate such as Karl Polanyi advocated in favor of democratic planning. After the stagnation and then the disintegration of the Soviet-style economies after 1989, (neo)liberal economics became triumphant. Yet, the economic meltdown of 2008, the increasing inequalities since the 1980s, and the ongoing environmental destruction under the dominant neoliberal economic regime, are requiring a revival of this debate [43].

Since 1989, it seems clear that the further oligarchization and monopolization of capitalist economies has demonstrated that a large amount of central planning takes place within multinational corporations and their complex global supply chains [44]; but importantly open source and "sharing" economies show there is also steep growth in nonmarket coordination, and the massive and complex technology projects through signaling, or stigmergy, whereby thousands of individuals and hundreds of corporate or noncorporate entities collaborate in open and transparent systems. It also seems that nation-states and classic multilateralism do not seem capable of adapting to the changes required to combat climate change and address resource scarcities.

In our 2019 report, *P2P Accounting for Planetary Survival* [17] we attempt a synthesis of the three major forms of economic coordination. We propose a base level of signal-based coordination for the production of the immaterial infrastructures, i.e., the collaborative production of software, design, and knowledge, through collaborative ecosystems. Once we have distributed ledgers and shared accounting and cooperative supply chains, the level of nonmarket coordination can also move to stigmergic cooperation in material production, including avoiding oversupply through a common vision of the flows of products and their consumption. Contributive accounting is the form of accounting, which can reward the types of contributions in this sphere.

The second level is for the human labor, capital and market goods that are in limited supply, and which can be traded and exchanged. Generative market mechanisms, using asset-based cryptocurrencies, currencies that reflect real material constraints, such as SolarCoin or FishCoin, are potential tools here. Mutual credit currencies will likely also play an important role, such as shown by the WIR/Sardex[8] models and the modern barter systems that already function for many businesses in the world. REA or flow accounting is the main tool in this context.

The third level is indicative, but also "coercive" planning. There are now various forms of thermodynamic and metabolic accounting schemes,

---

[8]See https://wiki.p2pfoundation.net/WIR_Economic_Circle_Cooperative; https://wiki.p2p foundation.net/Sardex

which can serve this purpose. For example, the R30[9] initiative has created a "Global Thresholds and Allocations Council" which is a scientific body which determines the flows and stocks of renewable and nonrenewable resources[10]. This global view can be integrated in regional/bioregional levels of production, and be transferred to the production entities operating in the region. This type of accounting shows the maximum level of resource usage that is permitted and organizes circular finance schemes to compensate for temporary overuse. According to the French engineer Francois Grosse [45], the growth of material usage cannot exceed 1% growth per year, lest it become exponential and would void any progress through circular economy models[11]. We speculate that new multilateral bodies would monitor these usage and taxation aspects. This level is also related to new forms of impact accounting, such as Compta Care[12] or the Economy of the Common Good[13] scheme [46]. The latter has 17 areas of impact, and argues that state bodies should start making taxation levels dependent on measured positive and negative impacts, changing the entire incentive structure of our economy to generative logics.

14.8 Conclusion

It this chapter, we have explored how blockchain and other technological tools may consider the social and ecological externalities in the transition toward cooperative, generative, and circular economies. DLT and related technologies that automate governance are rapidly expanding into, reshaping, and constructing new areas of social life. Uncritical design is likely to

---

[9]See https://www.r3-0.org/

[10]"The Reporting 3.0 Platform proposes the formation of a multistakeholder Global Thresholds & Allocations Council (GTAC), to establish an authoritative approach to reporting economic, environmental, and social performance in relation to generally accepted boundaries and limits." Originally found at https://drive.google.com/file/d/0B-9lY-KOycJ9M2NESU9rdzNqdU0/view

[11]Discussed by Christian Arnsperger, https://carnsperger.wordpress.com/2016/06/22/one-engineers-deeper-wisdom-francois-grosse-and-the-rediscovery-of-the-perma-circular-mindset/

[12]Details in French, at http://www.compta-durable.com/comptabilite-environnementale-sociale/modele-care/

[13]See here, https://wiki.p2pfoundation.net/Economy_for_the_Common_Good; information about the balance sheet model is at https://www.ecogood.org/en/ecg-balance-sheet/what-common-good-balance-sheet

reproduce structural inequalities reproduced in the under-recognition of service work or "care work" that enables organizations to function. Scholars have used a range of concepts to describe the under-acknowledged, under-rewarded, and often racialized and feminized labor in the context/processes of economic production. These include emotional and affective labor [47, 48, 49], immaterial labor [50, 51], and shadow work [52, 53]. While these dynamics were noticed well before the construction of the digital economy, they appear especially relevant in the context of digital platforms that derive economic value from unpaid contributions by their users [54, 51], including peer production projects [55, 56]. Entrepreneurs, coders, activists, policymakers, and other key actors who seek to effectively democratize and channel these technologies toward equitable outcomes need to be aware of how value decisions and material interests affect material design and the logic of the consequential infrastructures that derive from them.

Who will encode justice in the organizational design of the new economy? This is a question of governance, understood as the rules, policies, and procedures structuring the operation of an organization to implement decisions toward shared goals. There are also several important shifts that need to take place in the forms of value accounting that will be shared over these open and distributed ledgers. First, these digital value accounting systems must be fully open, transparent, and shareable when the actors need it. Accounting is what allows for mutual coordination mechanisms, now already the de facto standard for the open source production of code, design, and knowledge, to become the practice for physical production. In short, every transaction in the physical world will have its digital representation, and if these representations are shared, then actors in the physical production process can adjust their actions and processes to each other. But these actions based on the free coordination of agents in a collaborative system need to happen in a context of planetary boundaries. The threefold system we describe and propose in this chapter, offers the technical infrastructure to do just that. While it is a vision of a future infrastructure, we hope to have demonstrated that there are prototypes already in various stages of development, but they need, amongst other things such as funding, an integrative vision of how they might work together. Offering that vision has been our goal here.

Acknowledgments

We are grateful to Jeff Emmett for his explanations about the Commons Stack. Gratitude to Alex Pazaites for contributing as a coauthor to the original research for the P2P Foundation report, P2P Accounting for Planetary Survival, which helped us start this line of inquiry.

References

[1] Benkler, Y. (2006). "The Wealth of Networks". New Haven, CT: Yale University Press.

[2] Weber, S. (2004). "The Success of Open Source". Harvard University Press.

[3] Computer & Communications Industry Association (2011). "Fair Use in the U.S. Economy". Available online at http://cdn.ccianet.org/wp-content/uploads/library/CCIA-FairUseintheUSEconomy-2011.pdf (accessed December 31, 2019).

[4] Scholz, T. (2016). "Platform Cooperativism. Challenging the Corporate Sharing Economy". Ehmsen, S. and Scharenberg, A. (eds.). New York, NY: Rosa Luxemburg Stiftung.

[5] Bauwens, M. and Niaros, V. (2017). "Changing Societies through Urban Commons Transitions". P2P Foundation and Heinrich Böll Foundation. Available online at https://wiki.p2pfoundation.net/Changing_Socie ties_through_Urban_Commons_Transitions (accessed December 31, 2019).

[6] Bauwens, M. and Onzia, Y. (2017). "Commons Transition Plan for the City of Ghent" (in Dutch). Commissioned by the city of Ghent. English version: https://stad.gent/ghent-international/city-policy-and-structure/ghent-commons-city/commons-transition-plan-ghent

[7] Bauwens, M. (2005). "Political Economy of Peer Production". Available online at http://www.ctheory.net/articles.aspx?id=499 (accessed December 31, 2019).

[8] Heylighen, F. (2015). "Stigmergy as a universal coordination mechanism: components, varieties and applications", in T. Lewis and L. Marsh (eds.), "Human Stigmergy: Theoretical Developments and New Applications, Studies in Applied Philosophy, Epistemology and Rational Ethics". New York, NY: Springer. Available online at http://pespmc1.vub.ac.be/papers/stigmergy-varieties.pdf (accessed December 31, 2019).

[9] Waters-Lynch, J. (2018). "A theory of coworking: entrepreneurial communities, immaterial commons and working futures". PhD Thesis, RMIT University. Available online at https://researchbank.rmit.edu.au /view/rmit:162442 (accessed January 27, 2020).

[10] Salati, L.V. and Focardi, G. (2018). "The rise of community economy: from coworking spaces to the multi-factory model". Sarajevo: Udruženje Akcija. Available online at https://www.academia.edu/41043 179/THE_RISE_OF_COMMUNITY_ECONOMY_From_Coworking _Spaces_to_the_Multifactory_Model

[11] Kelly, M. (2012). "Owning Our Future: The Emerging Ownership Revolution. Journeys to a Generative Economy". San Francisco, CA: Berrett-Koehler Publishers. http://www.OwningOurFuture.com

[12] https://wiki.p2pfoundation.net/FLOSS_Foundations

[13] Arvidsson, A. (2019). "Capitalism and the Commons. Theory, Culture & Society". Available online at https://www.academia.edu/4023128 0/CAPITALISM_AND_THE_COMMONS? (accessed December 31, 2019).

[14] Arvidsson. A, (2019). "Changemakers. The Industrious Future of the Digital Economy". Stafford BC, Australia: Polity. Available online at http://politybooks.com/bookdetail/?isbn=9781509538898

[15] De Moor, T. (2008). "The silent revolution: A new perspective on the emergence of commons, guilds, and other forms of corporate collective action in Western Europe". IRSH, 53, Supplement, 179-212. Available online at https://www.ris.uu.nl/ws/files/20096187/_PUB_SilentRevolut ion_IRSH_53_Suppl.pdf

[16] Tarrius, A. (2015). "Étrangers de passage. Poor to poor, peer to peer". La Tour d'Aigues, L'Aube, 175.

[17] Bauwens, M. and Pazaitis, A. (2019). "P2P Accounting for Planetary Survival". Supported by the Guerilla Foundation and Shoepflin Foundation. Available online at http://commonstransition.org/wp-content/uploads/2019/09/AccountingForPlanetarySurvival_defx-2.pdf

[18] Kostakis, V. and Bauwens, M. (2014). "Network Society and Future Scenarios for a Collaborative Economy". New York, NY: Springer.

[19] Bauwens, M., Kostakis, V. and Pazaitis, A. (2019) "Peer to Peer. The Commons Manifesto". University of Westminster Press. London. DOI: https://doi.org/10.16997/book33

[20] Nakamoto, S. (2008). "Re: Bitcoin P2P e-Cash Paper". Available online at http://www.mail-archive.com/cryptography@metzdowd.com/msg1 0001.html (accessed April 4, 2018).

[21] Scott, B. (2014). "Visions of a Techno-Leviathan: The Politics of the Bitcoin Blockchain". E-International Relations. Available online at https://www.e-ir.info/2014/06/01/visions-of-a-techno-leviathan-the-politics-of-the-bitcoin-blockchain/ (accessed January 27, 2020).

[22] Noy, F. and Holemans, D. (2016). "Burgercollectieven in kaart gebracht". Oikos. Available online at http://www.collective-action.info/sites/default/files/webmaster/_PUB_Burgercollectieven-in-kaart-gebracht.pdf

[23] Iaione, C., Bauwens, M., Foster, S., et al. (2017). "The Co-Cities Report: building a "Co-Cities Index" to measure the implementation of the EU and UN Urban Agenda". LabGov & P2P Foundation. Available online at http://labgov.city/wp-content/uploads/sites/19/Co-cities-Open-Book-Report.pdf

[24] https://www.academia.edu/34360760/Urban_Commoning_Against_City_Divided_Field_Notes_from_Hong_Kong_and_Taipei?email_work_card=view-paper

[25] Manzini, E. (2013). "Small, Local, Open and Connected: Resilient Systems and Sustainable Qualities". Design Observer. Available online at https://designobserver.com/feature/small-local-open-and-connected-resilient-systems-and-sustainable-qualities/37670

[26] Oppenheimer, F. (2014). "Die Siedlungsgenossenschaft". Leipzig: Duncker/Humblot. Palgi, M. and Reinharz, S. (eds.).

[27] Sensorica (2019). Homepage. Available online at http://www.sensorica.co/ (accessed February 1, 2019).

[28] Zuboff, S. (2019). "The Age of Surveillance Capitalism: The Fight for a Human Future at the New Frontier of Power". New York, NY: PublicAffairs Hachette Book Group.

[29] Manski, S.G. (2017). "Building the blockchain world: Technological commonwealth or just more of the same?" Strategic Change, 26(5).

[30] Pickering, A. (1995). "The Mangle of Practice: Time, Agency, and Science". Chicago, IL: University of Chicago Press.

[31] Kaptelinin, V. and Nardi, B. (2006). "Acting With Technology: Activity Theory and Interaction Design". Cambridge, MA: MIT Press.

[32] Leonardi, P. (2012). "Materiality, sociomateriality, and socio-technical systems: what do these terms mean? How are they different? Do we need them?", in "Materiality and Organizing: Social Interaction in a Technological World", P. Leonardi, A. Nardi and J. Kallinikos (eds.). New York, NY: Oxford University Press, pp. 25–48.

[33] Robey, D., Raymond, B. and Anderson, C. (2012). "Theorizing information technology as a material artifact in information systems research", in "Materiality and Organizing: Social Interaction in a Technological World", Leonardi, P., Nardi, A. and Kallinikos, J. (eds.). New York, NY: Oxford University Press, pp. 217–236.

[34] Manski, S.G. and Manski, B.R. (2018). "No gods, no masters, no coders? The future of sovereignty in a blockchain world". Law & Critique, 29(2), 151-162. doi: 10.1007/s10978-018-9225-z

[35] Benkler, Y. and Nissenbaum, H. (2006). "Commons-based peer production and virtue". Journal of Political Philosophy, 14, 394-419. doi: 10.1111/j.1467-9760.2006.00235.x

[36] Winner, L. (1980). "Do Artifacts Have Politics?" Cambridgeshire: Daedalus, 121–136.

[37] Orlikowski, W.J. (2007). "Sociomaterial practices: exploring technology at work". Organization Studies, 28, 1435-1448. doi: 10.1177/0170840607081138

[38] Feenberg, A. (2012). "Questioning Technology". Abingdon: Routledge.

[39] Titcomb, A. (2019). "Crowdfunding the Commons". Noteworthy – The Journal Blog. Available online at https://blog.usejournal.com/crowdfunding-the-commons-d590238d8c3c (accessed January 27, 2020).

[40] Ostrom, E. (1990). "Governing the Commons". Cambridge, UK: Cambridge University Press.

[41] Brock, A. (2019). "CEPTR Core". Available online at http://ceptr.org/projects/core (accessed December 31, 2019).

[42] Morozov, E. (2019). "Digital Socialism. The Calculation Debate in the Age of Big Data". New Left Review, 116. Available online at https://newleftreview.org/issues/II116/articles/evgeny-morozov-digital-socialism

[43] Dyer-Witheford, N. (2013). "Red plenty platforms". Culture Machine, 14.

[44] Phillips, L. and Rozworski, M. (2019). "The People's Republic of Walmart". Brooklyn, NY: Verso.

[45] Grosse, F. (2011). "Quasi-Circular Growth: A Pragmatic Approach to Sustainability for Non-Renewable Material Resources". S.A.P.I.E.N.S., 2(4).

[46] Felber, C. (2015). "Change Everything: Creating an Economy for the Common Good". London, UK: Zed Books.

[47] Weeks, K. (2007). "Life within and against work: Affective labor, feminist critique, and post-Fordist politics". Ephemera: Theory and Politics in Organization, 7(1), 233-249.

[48] Federici, S. (2015). "From crisis to commons: Reproductive work, affective labor and technology in the transformation of everyday life. In "Psychology and the Conduct of Everyday Life", pp. 192-204. Abingdon: Routledge.

[49] Oksala, J. (2016). "Affective labor and feminist politics". Signs: Journal of Women in Culture and Society, 41(2), 281-303.

[50] Federici, S., (2008). "Precarious labor: A feminist viewpoint". In the Middle of a Whirlwind, available on-line at https://inthemiddleofthewhirlwind.wordpress.com/precarious-labor-a-feminist-viewpoint/

[51] Jarrett, K. (2014). "The relevance of "women's work" social reproduction and immaterial labor in digital media". Television & New Media, 15(1), 14-29.

[52] Illich, I. (1981). "Shadow Work". Salem, NH and London: Marion Boyars.

[53] Menzies, H. (1997). "Telework, shadow work: The privatization of work in the new digital economy". Studies in Political Economy, 53(1), 103-123.

[54] Terranova, T. (2000). "Free labor: Producing culture for the digital economy". Social Text, 18(2), 33-58.

[55] Restivo, M. and Van de Rijt, A. (2012)." Experimental study of informal rewards in peer production". PLOS ONE, 7(3).

[56] Arvidsson, A., Caliandro, A., Cossu, A., Deka, M., Gandini, A., Luise, V. and Anselmi, G. (2016). "Commons based peer production in the information economy". P2P Value. Available online at https://p2pvalue.eu/762-2/.

Index

About the Editors

Hans Schaffers is an independent researcher and advisor, focusing on human and societal implications of technological innovations. He received his PhD in Engineering Sciences at the University of Twente, The Netherlands, specializing in computer-assisted decision-making in large-scale water systems. He worked as an assistant professor in finance at the Erasmus University Rotterdam; as a senior researcher in technology and policy studies at the TNO; as a chief scientific consultant & and international affairs manager at the Telematica Instituut; as a visiting professor and research director at the Centre for Knowledge and Innovation Research at the Aalto University School of Business in Helsinki, Finland; as a research professor in digital business innovation at the Saxion University of Applied Sciences; and until recently working as a coordinator for Science, Management, and Innovation projects at the Radboud University Nijmegen. He has a broad experience as a project and programme coordinator in various European research and innovation initiatives, focusing on topics such as smart technologies, collaborative working environments, collaborative networked organizations, user oriented open innovation, urban and regional development, future internet test beds and living labs, and cooperative networks in industrial innovation.
E-mail: hans.schaffers@gmail.com

Matti A. Vartiainen is a senior advisor and professor emeritus of Work and Organizational Psychology at the Department of Industrial Engineering and Management, Aalto University School of Science. His research focuses on organizational innovations, digital work, new ways of working (remote work, mobile and multi-locational work, and distributed teams and organizations), collaborative working platforms, knowledge building, future competencies, and reward management systems. He has published many articles, for example, in *AI and Society*; *Brain Research*; *Group & Organization Management*; *Human Learning*; *Information and Software Technology*; *International Journal of Human Factors and Ergonomics in Manufacturing*; *Journal of Management*; *Journal of Organizational and End User Computing*; *International Journal of Project Management*; *Journal of Knowledge Management*; *New Technology, Work and Employment*; *Performance Improvement*, and the *Journal of Workplace Learning, Information and Software Technology*. He has edited and authored the following books among others: Andriessen, J. H. Erik & Vartiainen, M. (Eds.) (2006) *Mobile Virtual Work: A New Paradigm?* Heidelberg: Springer; Vartiainen, M., Hakonen, M., Koivisto, S., Mannonen, P., Nieminen, M.P., Ruohomäki, V. & Vartola, A. (2007) *Distributed and Mobile Work – Places, People and Technology*. Tampere: Otatieto; and Vartiainen, M., Antoni, C., Baeten, X., Hakonen, N. & Thierry, H. (Eds.) (2008) *Reward Management – Facts and Trends in Europe*. Langerich: Pabst Science Publishers.
E-mail: matti.vartiainen@aalto.fi

Jacques Bus received his PhD in science and mathematics from the University of Amsterdam. He worked as a researcher for 12 years and subsequently as a research program manager for 5 years at the Centrum Wiskunde & Informatica (CWI) in Amsterdam, The Netherlands. From 1988, he worked at the European Commission in leading positions in various parts of the Research programs ESPRIT and ICT, including IT infrastructure, program management, and software engineering, and since 2004 in trust and security, as well as the establishment of the Security Theme in the 7th Framework Programme of the European Commission. Since 2010 he works as an independent advisor in various projects and on Trust, Security, Privacy, and Identity in the digital environment. From 2011 to 2014, he has been the Director of Business Development of the Privacy & Identity Lab (PI.lab). Since 2011, he is the cofounder and secretary general of the "Digital Enlightenment Forum", a non-profit association in the field of digitization and society (https://digitalenlightenment.org/).
E-mail: secgen@digitalenlightenment.org